THE ENCYCLOPEDIA OF
ANIMAL PREDATORS

THE ENCYCLOPEDIA OF
ANIMAL PREDATORS

Learn about Each Predator's Traits and Behaviors
Identify the Tracks and Signs of More Than 50 Predators
Protect Your Livestock, Poultry, and Pets

Janet Vorwald Dohner

Storey Publishing

The mission of Storey Publishing is to serve our customers by
publishing practical information that encourages
personal independence in harmony with the environment.

EDITED BY Deborah Burns
ART DIRECTION AND BOOK DESIGN BY Michaela Jebb
TEXT PRODUCTION BY Erin Dawson
INDEXED BY Nancy D. Wood

COVER PHOTOGRAPHY BY © Don Johnston_MA/Alamy Stock Photo, back (top middle); © Donald M. Jones/
Minden Pictures/Getty Images, front (top left); © Fred LaBounty/Alamy Stock Photo, back (top right); © John
Foxx/Getty Images, front (top right); © Juniors Bildarchiv/GmbH/Alamy Stock Photo, back (bottom right);
© Mike Lentz Photography/www.mikelentzphotography.com, back (top left & bottom left); © Robert McGouey/
Wildlife/Alamy Stock Photo, front (bottom)

INTERIOR PHOTOGRAPHY CREDITS appear on page 280.

ILLUSTRATIONS BY © Elayne Sears, 117, 242–255, and scat
MAPS, TRACKS, SILHOUETTES, AND GRAPHICS BY Ilona Sherratt

Storey Publishing
210 MASS MoCA Way
North Adams, MA 01247
storey.com

Printed in China by R.R. Donnelley
10 9 8 7 6 5 4 3 2 1

Library of Congress Cataloging-in-Publication Data
on file

Dedicated to the belief that, armed with knowledge, we can coexist with animal predators on our farms, on our ranches, in our backyards, and in the greater world we share.

Contents

Part I

Predators
in the
Modern
World

One of the fundamental relationships in nature is that of predator and prey. To feed her fledglings, an eagle swoops with speed and grace to snatch a rabbit on the run. A wolf pack cooperatively chases down an elk, and with that success the whole pack eats. We humans are the ultimate predators, killing both to eat and to survive when threatened by an animal.

In the modern world, many of us are somewhat removed from the predatory act, other than observing a cat catching a mouse. Others of us, however, might walk out in our fields on a beautiful morning to find a gruesomely slaughtered lamb or a pile of decapitated chickens. Even in that moment of great anger and grief, the reality of predator and prey is inescapable and basic. We can't live in a world without predators; therefore, we must learn to coexist with the wild hunters around us while protecting what we raise.

Coexistence

For a long time, humans believed we could exterminate all large predators and shape the earth as we saw fit. We have since found that predators, both large and small, are essential to the healthy functioning of the earth's ecosystem. We have learned to appreciate the beauty of wild animals and their lives. Many people now work to save animals threatened with extinction, not only because our world is healthier when it is biologically diverse but also because our lives would be less rich without these animals.

Between the two points of view — protecting our domestic animals and valuing nature and all of its inhabitants — lies coexistence. Coexistence is possible, and it begins with knowledge. Knowledge of our predators' behaviors and habits is essential. Knowledge arms us when we encounter a predator on a walk in the backcountry. With knowledge, we learn how to design and implement predator-friendly systems that protect both our stock and ourselves.

Some of the methods of predator protection are old, as ancient as the shepherd who watched his sheep with his guardian dogs. Others are new, as wildlife biologists help us understand the predators around us rather than succumb to old myths or prejudices. These methods may require as much or more effort than simply eliminating all the predators, but when we value a balanced and sustainable world, they are worth the effort.

Consumers of meat, milk, or eggs can come to value predators and coexistence as well, just as they learn more about the reality of the lives of the farmers and ranchers who provide them with food.

Is not the sky a father and the earth a mother, and are not all living things with feet or wings or roots their children?

— Black Elk (Oglala Sioux)

CHAPTER 1

The Predation Situation

When European explorers and colonists arrived in the Americas, the single word they most often used to describe the "new world" they encountered was *abundance* — an abundance of land, natural resources, and animal life.

Of course, they weren't the first to discover the New World, because native peoples had long occupied and used the land, plants, and animals, trading commodities among one another. They altered the landscape, created agricultural fields, burned grasslands and forests to keep them open for grazing for favored herbivores, and may have been responsible for overhunting the megafauna after the last Ice Age.

Following that post–Ice Age era, some cultures became nomadic hunter-gatherers while others formed permanent communities for fishing or farming. Living more sustainably with nature, native peoples generally met their resource needs without the destruction of diversity and balance that lay on the horizon.

Taming Nature

European settlement would usher in an era of reckless exploitation that, from our contemporary viewpoint, was a truly stunning destruction of wildlife. The Merriam's elk, passenger pigeon, Carolina parakeet, and Labrador duck all went extinct due to relentless hunting. It was an extremely close call for the American bison, reduced from a population of 60 million to just 300 animals by 1900. Herons and egrets died by the thousands, primarily for plumes to decorate ladies' hats. Even the white-tailed deer became exceedingly rare in the eastern United States due to human activities. Animals were killed for their fur, hide, or feathers, but none of those was the primary reason that the large predators were destroyed.

The reason was that these animals were regarded as dangerous. As their natural prey was decimated, they became an increasing threat to the colonists' livestock. England, Scotland, Ireland, and other European homelands of the colonists had been eradicating wolves and other large predators for centuries. Since these animals were already extinct in the United Kingdom, livestock raisers there had no need to protect their livestock from large or even medium-sized predators. With their only predator being the fox, which was widely hunted on foot and on horseback, sheep were turned loose to graze without active shepherding or livestock guardian dogs (LGDs). There was no tradition of protecting sheep or cattle from serious predation.

It is therefore not surprising that the wolves, bears, and mountain lions of the eastern colonies seemed terrifying. Taming nature was the first order of business. The first wolf bounty was set in 1630 in the Massachusetts Bay Colony. From that point forward, the colonies and later the states set bounties for killing the large predators. Thus the hunting, trapping, and poisoning commenced.

By 1870, no mountain lions remained in the eastern states or provinces. By the beginning of the 20th century, wolves were gone from the continental United States and the adjoining areas of Canada except the northernmost Great Lakes region and northern and western Canada. By the 1930s, wolves, mountain lions, and grizzly bears were nearly eliminated from the Intermountain West, except for isolated pockets.

The loss of the large predators and the changing landscape allowed the opportunistic coyote, originally a resident of the Great Plains and arid West, to expand its territory throughout the continent. The widespread coyote is now the single greatest threat to livestock and poultry raisers. The loss of large predators also allowed small predators — raccoons, opossums, and skunks — to increase in numbers and to expand their ranges as well.

▲ Wolf　　　　▲ Mountain Lion　　　　▲ Bear

▲ By the 1930s, the three large North American predators — wolves, mountain lions, and grizzly bears — were nearly eliminated from the continental United States.

▲ John James Audubon's illustrations brought viewers face-to-face with wild animals in their own element.

Efforts to Coexist

While the destruction proceeded, the foundations of the conservation movement were also being laid. The 19th century saw the origin of a number of progressive movements — to abolish slavery, fight for women's rights, establish laws regarding child labor and food and drug safety, and regulate animal welfare. Alongside this progressivism, the love of nature grew. With it came efforts to preserve and protect lands, animals, and natural resources. As the cities became increasingly crowded, people grew to value parks and recreation land for camping, hiking, and bird watching. Sympathy and empathy grew for wild animals.

This renaissance was fostered by a number of famous writers, artists, and naturalists. Some were both conservationists and hunters, which was not a contradiction. Others were explorers of the wilderness, writing and making images to reveal its natural beauty. Among the first and foremost was John James Audubon, an artist, naturalist, and the most famous ornithologist of his time. Other artists included George Catlin, Thomas Moran, Albert Bierstadt, and Frederic Edwin Church. John Burroughs, Henry David Thoreau, Ralph Waldo Emerson, John Muir, William Cullen Bryant, George Perkins March, Ernest Thompson Seton, and Theodore Roosevelt wrote about nature and wild animals. Theodore Roosevelt and the writer and naturalist George Bird Grinnell, founder of the Audubon Society (1905), could be considered the earliest conservationists.

The Conservation and Preservation Movements

In 1860, 80 percent of the US population still lived in rural areas, but the march toward urban development was proceeding. As Frederick Law Olmsted designed New York City's Central Park, efforts were beginning in the states to preserve land in the Adirondack and Catskill Mountains. Soon after biologist Ernst Haeckel first used the word *ecology*, John Wesley Powell, back from his exploration of the Colorado River, expressed the importance of recognizing the environmental fragility of the West. The world's first national park — Yellowstone — was established in 1872. By this time, the concept of *conservation* implied management of natural resources and public lands for human use, encompassing the forests, grazing land, minerals, water, fish, birds, and other animals. Sustainable and responsible use of this land was essential to conservation.

The turn of the century saw the election of Theodore Roosevelt as president. A lover of hunting and the hunting lifestyle, he worked for the conservation of game species and the environment they needed. Hunters such as Roosevelt believed it was not sporting to kill predator species. Predators should be killed only to maintain game animals and birds for hunting. Roosevelt and fellow conservationists such as Gifford Pinchot worked to create and protect national forests, wildlife refuges, and parks.

At the same time, the separate concept of *preservation* was taking root. John Muir founded the Sierra Club in 1892 with the mission to preserve and protect wilderness. Preservationists wanted humans to value wild land and animals solely for their natural beauty and not their usefulness to humans. One of the earliest conflicts between these two points of view was the conservationists' proposed dam in the Hetch Hetchy Valley of Yosemite National Park. Flooding the valley created a huge water supply for San Francisco, while also destroying the beauty and life found in the valley. From that point on, tension and conflicting priorities increased between conservationists and preservationists.

Game and Land Management

While conservation and preservation were gaining ground, predator control efforts continued. The National Animal Control Act for the destruction or control of predators was passed in 1931 (although Congress had already been providing funds for wolf and coyote control during the preceding 15 years). Famed conservationist Aldo Leopold developed game management science in the model of forestry science. By the 1940s, he became deeply concerned that the burgeoning population of deer and elk, no longer controlled by predators, was threatening the environment.

In the 1930s, in the days of the Dust Bowl, evidence mounted that unregulated land use was destructive. The Taylor Grazing Act was passed to "stop injury to the public lands by preventing over-grazing and soil deterioration." In 1946, the Bureau of Land Management, now in the Department of the Interior, replaced the earlier Grazing Service. Early water and air pollution acts

laid the groundwork for future ecological regulation. In 1949, Aldo Leopold wrote *A Sand Country Almanac*, which is widely regarded as one of most influential environmental books of the 20th century. By 1951, the Nature Conservancy was founded, an organization that would come to own and manage the largest collection of private nature reserves in world.

▲ "Wilderness is not only a haven for native plants and animals but it is also a refuge from society. It is a place to go to hear the wind and little else, see the stars and the galaxies, smell the pine trees, feel the cold water, touch the sky and the ground at the same time, listen to coyotes, eat the fresh snow, walk across the desert sands, and realize why it's good to go outside of the city and the suburbs."

— JOHN MUIR, 1901

"We reached the old wolf in time to watch a fierce green fire dying in her eyes. I realized then, and have known ever since, that there was something new to me in those eyes — something known only to her and to the mountain. I was young then, and full of trigger-itch; I thought that because fewer wolves meant more deer, that no wolves would mean hunters' paradise. But after seeing the green fire die, I sensed that neither the wolf nor the mountain agreed with such a view. . . .

"I now suspect that just as a deer herd lives in mortal fear of its wolves, so does a mountain live in mortal fear of its deer. And perhaps with better cause, for while a buck pulled down by wolves can be replaced in two or three years, a range pulled down by too many deer may fail of replacement in as many decades. So also with cows. The cowman who cleans his range of wolves does not realize that he is taking over the wolf's job of trimming the herd to fit the range. He has not learned to think like a mountain. Hence we have dustbowls, and rivers washing the future into the sea."

— ALDO LEOPOLD, 1949

Environmental Awareness Grows

In the 1960s, growing concerns and serious problems came to the fore with a resulting flurry of government action. Just as the photography of Ansel Adams was capturing the ethereal beauty of nature, two important writers captured the crisis the country faced. Both Rachel Carson, who wrote *Silent Spring* in 1962, and Stewart Udall, with *The Quiet Crisis* in 1963, brought widespread public awareness to a broad and deepening environmental situation.

As secretary of the interior, Udall championed a series of important pieces of conservation legislation: the Wilderness Act, the Endangered Species Preservation Act (later expanded as the Endangered Species Act), the National Trails Act, and the Wild and Scenic Rivers Act. Other important environmental efforts included the National Environmental Quality Act, the Clean Air and Clean Water Acts, and the Environmental Policy Act. The United States became the first nation to pursue far-reaching environmental conservation efforts. In 1992, the Convention of Biological Diversity convened in Rio de Janeiro as part of the Rio Earth Summit and gathered worldwide resolve to promote sustainable development for the needs of people, animals, and plants; in 2002, Environment and Climate Change Canada enacted the Species at Risk Act of Canada (SARA) to protect wildlife. In this book, when the terms Endangered and Threatened are capitalized, they refer to species listed under the Endangered Species Act (www.fws.gov/endangered/species/us-species).

Among the very first animals listed by the US government as Endangered species were the gray wolf, the Florida panther, and the grizzly bear. For the first time, predators were offered official protection. With this move, the philosophies and methods of preservation and conservation began to come together in a new approach.

Predator Conservation Concepts and Terms

A **keystone species**, even though small in number, may have a large effect on the environment.

The **apex predator** sits at the top of the food chain. An adult apex predator is not prey for other animals except humans, who are also an apex predator.

Habituation is an adaptive behavior in which the predator grows used to human presence and stops responding appropriately to it, leading to potentially dangerous interactions.

Fragmentation occurs when populations of a given species are geographically separated by human development.

A **wildlife corridor** is a connecting area between habitats to foster genetic diversity among individuals in a species that may have become fragmented.

Rewilding is conservation on a large scale, including protecting or reintroducing apex predators or other keystone species, restoring and protecting wilderness and habitat, and providing corridors to promote genetic diversity.

Agencies and Acts

BLM Bureau of Land Management

CDC Centers for Disease Control and Prevention

CITES Convention on International Trade in Endangered Species of Wild Fauna and Flora

COSEWIC Committee on the Status of Endangered Wildlife in Canada

DNR Department of Natural Resources (for individual states)

ESA Endangered Species Act

FWS US Fish and Wildlife Service (manages fish, wildlife, and natural habitats; enforces federal wildlife laws; protects endangered species; conserves and restores wildlife habitat)

IUCN International Union for Conservation of Nature

SARA Species at Risk Act of Canada

USDA US Department of Agriculture

USFS US Forest Service

Conservation Biology

First presented as a concept in 1978, conservation biology is the scientific study of nature and biodiversity in order to preserve species, habitats, and ecosystems. Conservation biology seeks to bring together ecology, which studies organisms and their environment, and conservation, in both policy and practice. It recognizes the importance of carnivores in helping to maintain biodiversity, keep prey species healthy, and hold their numbers in balance.

Public Health Issues

Predation on domestic animals isn't the only danger from proximity to wild animals. Besides the contamination of crops and feedstuffs by smaller predators, various diseases can be passed from animals to humans, caused by bacteria, parasites, or viruses. These zoonotic diseases can have serious or fatal consequences. Water can become contaminated with giardia, leptospirosis, or salmonella, and dogs, cats, and other livestock can acquire these diseases from drinking it.

Rabies

Wild animals are responsible for 92 percent of all reported rabies cases in the United States and Canada; domestic animals account for the rest. In both countries the primary carriers are raccoons, bats, skunks, foxes, and, less often, coyotes — and the most common carrier varies by state or province. Rodents and rabbits rarely contract rabies, and opossums are rabies resistant. After an encounter, an increasing number of people take postexposure treatment. Only 1 to 3 fatal cases of rabies occur in humans each year in the United States, however, and the disease is even rarer in Canada. Most fatalities occur because the individuals did not know they were exposed or failed to receive appropriate and timely treatment.

Rabies is a viral disease affecting the central nervous system. The virus is transmitted through the saliva of an infected animal, usually by a bite, exposure

▲ Raccoons remain the most frequently reported rabid wildlife, followed by bats, skunks, and foxes. Detailed information, by state, is available on the CDC website for both domestic and wildlife rabies cases.

to an open wound, or contact with mucous membranes. If your dog or another animal has encountered or fought with a potentially rabid animal, use extra caution to avoid coming into contact with any fresh saliva on its coat. The virus cannot survive after the saliva has completely dried. Wear protective gloves if you bathe your dog after an encounter.

Symptoms. Rabies in wild animals can include one or more symptoms: agitation, biting or snapping and drooling, unnatural tameness or lack of fear, loss of coordination, wobbly or circling behavior, paralysis, disorientation, self-mutilation, or daytime activity in a nocturnal animal.

Treatment. Medical attention is essential for all wild animal bites to humans, pets, and livestock. Before seeking emergency treatment for a bite from a potentially rabid animal, immediately and thoroughly scrub the wound with antiseptic soap or Betadine and rinse extensively. Treatment, when received in a timely manner, is 100 percent effective. All dogs and cats that come in contact with wildlife should be vaccinated against rabies as a preventive. Contact the authorities if you capture or suspect a rabid animal.

Other Diseases

In addition to rabies, other serious zoonotic diseases include the following:

- **Baylisascaris.** Roundworm transmitted through feces or soil contaminated by raccoon, bear, or skunk feces

- **Brucellosis.** Bacteria possibly transmitted through handling feral pig carcasses

- **Echinococcosis.** Tapeworm found in scat (feces) of foxes, coyotes, cats, and other canine or feline predators

- **Hantavirus.** Virus transmitted through dust from rodent urine, scat, or nest material

- **Lymphocytic choriomeningitis virus (LCMV).** Virus transmitted by rodents, feral hogs, skunks, and raccoons through urine, scat, or saliva

- **Plague.** Bacteria transmitted from infected fleas from rodents (Cats or other animals may get the plague from eating infected rodents.)

- **Rat-bite fever.** Bacteria transmitted by bites or scratches and by water or food contaminated by urine

- **Toxoplasmosis.** A parasite shed in cat feces

- **Tuberculosis.** Bacteria transmitted by opossums or badgers

For more information on these diseases and others, contact the US Centers for Disease Control and Prevention at cdc.gov/healthypets/pets/wildlife.html.

▶ Although any mammal may be infected with rabies, some carnivore species are major reservoirs for the virus in distinct variants. Bats may also serve as rabies reservoirs.

Skunk

Raccoon

Fox

Skunk and fox

Knowledge Is Power

For many reasons, we must learn about predators and their habits and behaviors, how to identify them, and how to prevent damage. Ultimately, we must learn to coexist in the world we share, understanding the value and place of predators in the ecosystem. Some predators actually do pose a threat to people, although a small one. We may have forgotten how to live with predators around our homes and wild places, so we must learn how to avoid risky behaviors, reduce attractants, and be proactive in preventive techniques and strategies.

The bottom line: We don't need to fear predators, but we do need to learn how to deal with them — whether in our chicken coops, our pastures, or our own backyards.

Wildlife Handling Guidelines

- Do not pet or touch wildlife.

- Wear gloves when handling dead animals, cleaning up feces, gardening, or working with potentially contaminated animal feedstuffs.

- Dispose of feces or nest materials in a safe manner. Dry feces can spread disease through airborne particles.

- Disinfect contaminated areas, especially outdoor food preparation areas and tools.

- Wash hands thoroughly after gardening, working outdoors, or handling dead animals or feces.

- Clean any bites or scratches thoroughly and seek medical attention.

- Keep outdoor sandboxes covered.

- Wash fruits and vegetables that may have been in contact with infected soil.

- If you find an injured, orphaned, or sick wild animal, call the appropriate authorities. To find a wildlife rehabilitator in the United States or Canada, consult the National Wildlife Rehabilitators Association at nwrawildlife.org.

- Don't rescue wild or sick animals. Call authorized rehabilitators or animal control.

CHAPTER 2

Who's Out There?

When a livestock or poultry owner discovers the carcass of an animal, it can be emotionally and financially devastating. Owners ask themselves what they could have done differently to protect their animals. Protection begins with an understanding of what threats exist in the place where you and your animals live.

Three factors help to determine your potential threats: the region where you live; your home's location, from city to suburb to semirural exurb to truly rural area; and finally, the animals you keep, whether large stock or small animals, poultry, or pets. Learning about predator ranges, livestock death statistics, and the migration of predators into suburban and urban environments informs farmers and homeowners alike.

Identifying Likely Culprits

The US Department of Agriculture (USDA) tracks adult cattle, calf, adult sheep, and lamb losses through two major reports, with more limited information collected for goats (see box on page 14). This information helps rural and exurban residents determine their predator threats, judge their relative risks according to area of the country, and evaluate the success of various nonlethal predator control methods. It also clearly illustrates the scope and impact of predation on livestock and their owners.

By far, most cattle and sheep are lost in ways that do not involve predators. Predation is responsible for 4 to 6 percent of all cattle and sheep deaths.

- 1.8% of adult sheep lost to predation

- 3.9% of lambs lost to predation

- 2.3% of adult cattle lost to predation

- 8% of calves lost to predation

Other causes of livestock losses include old age, lambing problems, disease, parasites, weather, accidents, poisoning, and theft.

Goat losses are significantly higher, at more than 30 percent of all deaths, a fact that certainly deserves more examination. Coyotes are the foremost predator of goats, as reported by many sources, although a USDA breakdown of predator-caused goat deaths isn't available.

Overview of Predator Threats

The USDA figures also show us how small and large operations differ and how the predator threats vary across regions of the country.

Sheep. In flocks with more than 25 animals, the coyote is the primary killer, while in smaller farm flocks, domestic dogs are the greatest threat. In pastured or range flocks larger than 1,000 sheep, coyotes remain the major predator by far, followed by mountain lions, bears, and wolves. Bears kill more sheep in Colorado, Utah, and Nevada. Dogs cause more deaths than coyotes do in New Mexico.

Cattle. In cattle herds, coyotes are responsible for about 60 percent of all predator losses, but small operations of less than 50 animals lose twice as many animals to predators than do larger operations. The southwestern states have the highest percentage of adult cattle loss to predators, followed by southeastern areas; and the Southeast has experienced the greatest increase in calf losses. The fact that coyotes have recently moved into the Southeast is no doubt very relevant to the situation.

Goats. Almost half of all the goats in the United States are raised in Texas. Across the country there are about 2 million meat goats and 0.5 million dairy and fiber goats. Predator losses are high, especially in the Southeast and Texas.

Intersection of Rural and Urban

Rural areas can vary from wilderness and open range to recreational and suburban use; however, dealing with predators is not just a rural problem by any measure. The altered landscape and the loss of the apex predators (wolves, mountain lions, and grizzly bears) opened the door to the other predators that have adapted and flourished — black bears, coyotes, bobcats, foxes, raccoons, and skunks. These predators have greatly expanded their ranges and numbers in recent decades.

Meanwhile, as housing and recreation have pushed into rural and undeveloped areas, large-predator threats are growing, from coyotes to the large cats, bears, and wolves. At this intersection of rural and suburban, on both small and hobby farms, we now see many such threats to small livestock, poultry, pets, and people. Predators are creeping into backyards and neighborhoods, attracted by bird feeding, garbage, rodents, and feral cats.

Sheep and Lamb Loss Due to Predators

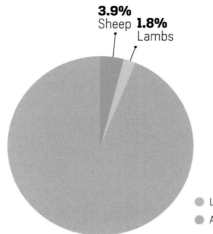

3.9%
Sheep **1.8%**
Lambs

- ○ Losses due to predators
- ● All causes of death

In a typical year, **less than 6% of sheep and lambs are killed by predators.**

▶ Other causes of death include old age, lambing problems, disease, parasites, weather, accidents, poisoning, and theft.

▶ The #1 cause of sheep loss is old age at 71.9%.

▶ The #1 cause of lamb loss is weather-related causes at 63.6%.

Sheep and Lamb Loss by Specific Predator

PREDATOR	SHEEP	LAMB
Coyote	54.3%	63.7%
Dog	21.4%	10.3%
Mountain lion	5.6%	4.5%
Bear	5.0%	3.0%
Wolf	1.3%	0.4%
Bobcat/lynx	1.1%	2.8%
OTHER PREDATORS		
Black vulture	0.6%	1.4%
Fox	0.5%	1.9%
Feral pig	0.4%	0.7%
Eagle	0.2%	3.7%
Other animals	0.6%	0.6%
Unknown predator	8.6%	5.5%

USDA, September 2015.

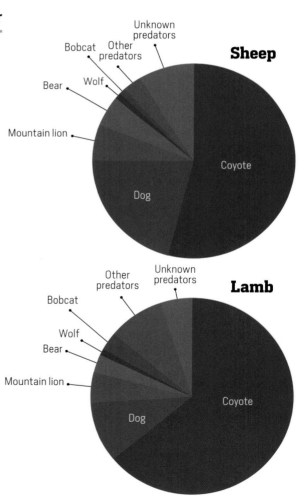

Assessing Your Predator Risks

As a first step, homeowners, farmers, and ranchers need to arm themselves with knowledge. Specific predator information relevant to your area can be found from the natural resource departments in states and provinces, the USDA extension offices, the Ministry of Agriculture in Canada, and livestock producers' organizations. The USDA reports on death loss provide further specifics state by state. Stay informed on the current situation by talking to neighbors and following local news stories. Your observations are essential, not only to assess which predator is present but also to pinpoint the time of year and location on your property. Your observations need to be further informed by serious investigation of any predator attacks.

To evaluate your risk of a predator encounter or attack, consider the following:

- **Predation statistics** from the US National Agricultural Statistics Service or Canada's Ministry of Agriculture, to determine most likely threats

- **Predator identification** in your local area, even at your specific farm or home

- **Predator pressure changes** during the year, including offspring needing food, and weather or season reducing access to prey animals

- **Predator movements** that are likely in the future as animals migrate into new areas

- **Predator attractions,** including your terrain, stock, and husbandry

---

Beware of the Dogs

Clearly, coyotes are the primary predator that livestock raisers face across the continent. In many areas, they are a relatively new threat. The startling revelation, however, is that domestic dogs kill the next-largest number of livestock. Dogs also cause the most emotional upset, because they slaughter animals viciously, in large numbers — and not primarily for food.

Removing free-roaming dogs, most of which are not feral, from the predation statistics would make these charts more accurately represent the threat that true wild predators pose for livestock.

As you analyze your particular situation, pay attention to the terrain as it relates to your husbandry practices.

- Heavily wooded, rough, or very large areas are riskiest for stock.

- Many predators are reluctant to cross large open areas.

- Grazing areas farther removed from human residence are more vulnerable.

- Land with a mosaic of fields and forests and areas with high deer population are attractive to predators.

- Unprotected hay or other crop storage will attract deer and elk, and their predators.

Predator Threat by Region

This detailed information from the USDA will help determine what is likely to be a threat in your region.

Percentage of Cattle Loss by Predator by Region

PREDATOR	NC	NE	NW	SC	SE	SW
Coyote	64.2	47	17.6	30.5	49.9	24.4
Dog	4.7	7.0	0.3	17	21.5	1.7
Mountain lion, bobcat/lynx	20.2	1.6	4.4	16.3	0.2	28.1
Other predators	6.1	20.6	31.8	13.4	12.5	11.6
Unknown	4.8	23.8	45.9	22.8	15.9	34.2

Percentage of Calf Loss by Predator by Region

PREDATOR	NC	NE	NW	SC	SE	SW
Coyote	71.7	73.6	48.5	42.8	65.8	66.5
Dog	2.1	6.0	1.2	14	12.6	5.9
Mountain lion, bobcat/lynx	9.8	0.5	6.8	12.9	0.4	12.5
Other predators	10.2	12.3	29.2	18.2	13.8	7.8
Unknown	6.2	7.6	14.3	12.1	7.4	7.3

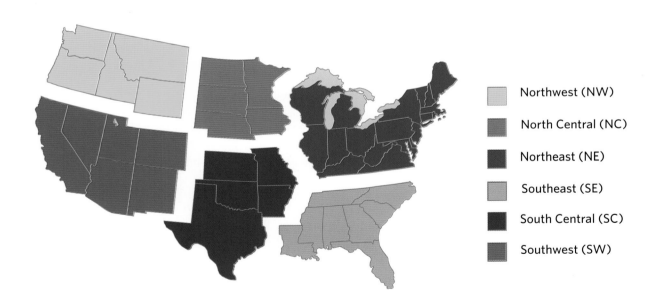

- Northwest (NW)
- North Central (NC)
- Northeast (NE)
- Southeast (SE)
- South Central (SC)
- Southwest (SW)

Was This an Attack?

Your own observations about predators on your property are essential. When you find a dead animal, you need to determine if it died from natural causes or a predator attack. In most cases animals and poultry die of natural causes rather than a predator attack. Your ability to determine the cause of death is partially determined by how rapidly you find the carcass before scavengers show up and begin to feed.

Try not to disturb the physical evidence of tracks, scat, blood on the ground, drag marks, disturbed vegetation, or a covered carcass. Take photographs, if possible, and record your observations. Wear gloves and protective clothing when examining a dead animal or afterbirth. In many cases, assistance from a veterinarian or wildlife damage expert can be valuable in determining natural causes of death versus a predator attack — especially if you are experiencing multiple deaths.

Questions to consider:

Can you determine the time the attack occurred?

Are there signs of a struggle? Signs may include torn wool, hair, or feathers; blood splatter; drag marks; and damaged vegetation.

How are your other animals behaving? Unusual or unsettled behavior may include nervousness, scattering, and increased vocalization.

Are there bite marks? If so, where are they located and what is their size? Take measurements and photographs to help you identify the predator. Try to identify whether it was a mammal or a raptor. You may need to clip hair or wool to look for puncture marks. When bites are made to a living animal, there will be bruising or hemorrhaging under the skin.

Is there significant blood? Profuse bleeding occurs before death and for a short time after. A death from natural causes, not an accident or attack, may show a loss of bodily fluids such as urine but not much blood.

Is the animal a newborn or a stillborn? Stillborn animals may have soft membranes covering the hooves. A field autopsy can also reveal important information. Pink lungs indicate the animal was breathing before death, while stillborn animals have dark-colored lungs that will not float in water. Milk in the stomach indicates the newborn was able to nurse before death.

KEEPING AN EYE OUT

Trail cameras can help you determine which predators are nearby or threatening your stock, where they are testing or accessing your fences, and whether they have become habituated to other predator protection measures. The newer trail cameras have better resolution, increased memory, and better battery life than older models.

Nighttime flash can be infrared or LED. Other features to consider include the time delay or trigger speed between the motion sensor detecting an animal and the shutter being engaged, and the speed of consecutive photos. Sturdy cases can be very important if predators, especially bears, might investigate the cameras.

Poultry Damage ID Guide — Common Predators

After identifying potential culprits below, check individual profiles in Part II for additional details, observations, tracks, and scat.

Note: Predators can occasionally be active during nonnormal times or behave in atypical ways.

		Badger	Bobcat	Cat	Coyote	Crow	Dog	Eagle	Fisher	Fox	Hawk	Human	Lynx	Magpie	Marten	Mink	Mountain Lion	Opposum	Owl
Time	Day		✔	✔		✔	✔	✔		✔	✔	✔		✔					
	Dusk		✔	✔	✔	✔			✔	✔	✔	✔	✔		✔		✔		✔
	Night	✔	✔	✔	✔	✔			✔	✔		✔	✔		✔	✔	✔	✔	✔
	Dawn		✔	✔	✔	✔	✔	✔		✔		✔	✔		✔		✔		✔
Eggs	Eggs eaten					✔	✔			✔				✔	✔			✔	
	Eggs missing					✔						✔		✔				✔	
Missing Birds	One bird missing		✔	✔	✔			✔	✔	✔	✔			✔			✔	✔	
	Multiple birds missing											✔							
	One or more chicks missing				✔													✔	
Dead Birds (Partially or Mainly Eaten)	One or two birds killed	✔	✔	✔	✔			✔		✔	✔							✔	✔
	Several birds killed			✔			✔			✔				✔	✔				✔
	Chick or chicks killed			✔		✔							✔					✔	
	Heads removed or eaten		✔	✔						✔						✔		✔	✔
	Bodies piled together															✔			
	Missing limbs or heads pulled through fence						✔												
	Breast or anal area eaten								✔						✔			✔	
	Abdomen eaten															✔			
	Breast and legs eaten										✔								
Teeth or Talon Marks	Talon punctures on head or body							✔			✔								✔
	Bites on neck				✔												✔		
	Bite on head or neck		✔							✔			✔		✔				
	Many bites on head, neck, body, or legs															✔			

Raccoon	Rat	Raven	Skunk	Snake	Vulture, Black	Weasel
		✔		✔	✔	✔
✔	✔		✔	✔		✔
✔		✔	✔			✔
✔	✔	✔		✔		
	✔			✔		
✔			✔			
✔						✔
	✔	✔	✔			
✔						✔
						✔
✔						
			✔			
✔						
			✔			
			✔			✔

Other Observations

Event	Cause
Animal(s) killed and mauled but not eaten	Dog
Bites on legs of live birds	Rat
Dead chicks or birds stuck in tunnels	Rat
Musky smell	Skunk, weasel, mink
Feathers on ground	Fox, coyote, hawk, owl
Wounds or pulled feathers on back and tail of live bird	Cannibalization
Injuries on back, pulled feathers	Rooster mounting hen
Several dead birds piled against fence or in corners, carcasses flattened	Fright and panic due to chasing by dogs, wolves, or other larger predators
Serious damage to coop	Bear
Latches opened	Raccoon, human

Livestock Damage ID Guide — Likely Suspects

After identifying potential culprits below, check individual profiles in Part II for detailed information, observations, tracks, and scat. Small predators can kill, carry away, or consume only very small livestock — rabbits or small lambs, for example. Predators can occasionally be active during nonnormal times or behave in atypical ways.

	Badger	Bear	Bobcat	Cat	Coyote	Crow	Dog	Eagle	Feral Hog	Fisher	Fox	Lynx	Magpie	Marten	Mink	Mountain Lion	Raccoon	Rat	Raven	Skunk	Vulture, Black	Weasel	Wolf
Time — Day		✔	✔	✔		✔	✔	✔	✔		✔		✔						✔		✔	✔	
Time — Dusk		✔	✔	✔	✔		✔			✔	✔	✔		✔		✔							✔
Time — Night	✔	✔	✔	✔	✔		✔			✔	✔	✔		✔	✔	✔	✔	✔	✔		✔	✔	✔
Time — Dawn			✔	✔	✔		✔	✔			✔	✔		✔		✔							
Missing Animal		✔	✔	✔	✔		✔	✔			✔	✔									✔		
Dead Animal — One or two animals killed		✔	✔	✔	✔							✔				✔					✔		✔
Dead Animal — Several animals killed						✔	✔	✔		✔		✔		✔	✔	✔						✔	✔
Dead Animal — Animal killed not consumed				✔			✔																
Dead Animal — Small animal, only head and neck consumed				✔						✔				✔	✔						✔	✔	
Dead Animal — Rabbit or lamb consumed except head and fur	✔																						
Dead Animal — Small animal consumed		✔	✔					✔			✔					✔							
Dead Animal — Opened ventrally		✔	✔		✔						✔												
Dead Animal — Vital organs consumed first, not rumen and intestines (may be pulled out)		✔	✔		✔						✔					✔							✔
Dead Animal — Udder (consumed?)		✔																					✔
Dead Animal — Breast and neck eaten								✔						✔									
Dead Animal — Small wounds in body																✔						✔	
Dead Animal — Meaty areas consumed		✔	✔		✔						✔					✔							✔
Dead Animal — Newborn eyes, nose, tongue, genitals, rectum, or hooves pecked						✔							✔						✔		✔		
Dead Animal — Newborn, nose or other parts chewed																	✔	✔					
Dead Animal — Nose, lower jaw, ears, palate, brains eaten								✔															

		Badger	Bear	Bobcat	Cat	Coyote	Crow	Dog	Eagle	Feral Hog	Fisher	Fox	Lynx	Magpie	Marten	Mink	Mountain Lion	Raccoon	Rat	Raven	Skunk	Vulture, Black	Weasel	Wolf
Teeth, Talon, or Claw Marks	Small animals bitten through forehead, top of head or neck, or back		✔			✔				✔	✔	✔			✔	✔	✔						✔	✔
	Skull or neck crushed		✔							✔														
	Deep talon marks in head or neck								✔															
	Small animal bitten in throat			✔		✔						✔	✔											✔
	Animal killed or eaten through abdomen or rectum					✔					✔				✔									✔
	Larger animal bitten through top of neck or back	✔	✔										✔				✔							
	Larger animal bitten head, neck, back, flank, or hind					✔																		
	Larger animal bitten hind, sides, shoulders, tail, or nose																							✔
	Larger animal, deep talon marks in back or upper ribs								✔															
	Bites in side, rips, gashes					✔		✔																
	Claw marks and gashes on shoulder, back, flanks, body		✔	✔									✔				✔							
	Deep talon marks																							
	Clean edges on bones and flesh			✔		✔							✔											
	Large bones crushed																✔							
Other Observations	Carcass partially or completely skinned out		✔						✔	✔														
	Carcass cached a distance away	✔	✔									✔					✔							✔
	Vegetation crushed at site		✔							✔														
	Wounds on live animal					✔		✔																
	Mutilated animals; ears, tails, wool, or fur torn off; broken legs							✔																
	Extremely stressed and anxious survivors							✔																
	Strong smell		✔																		✔			

Part II

Predators
Up Close

The definition of a predator is quite simple: an animal that eats prey. All predators actively kill or scavenge another animal for food. Types of predators are classified as follows.

Hypercarnivores, such as members of the cat family, eagles, and snakes, must eat meat because of their digestive-system needs. Also referred to as **obligatory carnivores**, they must have some meat in their diet. These species engage in active pursuit or ambush behaviors.

We describe other animals as **omnivores**, which means they are able to fulfill their dietary needs from a variety of plant and animal sources. Many omnivores are **opportunistic predators**, which don't always actively hunt but do take advantage of prey or carcasses they encounter as they forage. Opportunistic predators will kill and eat a wide variety of prey, unlike other predators, which specialize on a narrow range of prey.

Some predators, such as wolves, use social cooperation to kill animals much larger than themselves. Other predators are solitary in their hunting behaviors.

An **apex predator** resides at the top of the food chain, meaning that as adults, members of that species are not preyed upon by any other predator.

Why We Need Predators

Predators are often described as **keystone species** for their essential role in ecology. Predation is a key element of ecosystems, preventing overpopulation of prey species that can be extremely destructive to the land. These wild hunters increase biodiversity by not allowing one species to dominate all others. Predators also maintain the health of a prey species by eliminating ill or weak members of the group.

When predators are eliminated or severely reduced through habitat loss, the entire ecosystem suffers. For example, when wolves were eliminated and landscapes changed, coyotes and other smaller predators were able to increase their populations and ranges. As an unintended consequence, the coyote is now the primary killer of domestic livestock in the United States.

When apex predators are restored, some biologists posit that a **trophic cascade** occurs from the top of the food pyramid. The herbivores that overgraze are reduced, plant ecosystems are restored, and other species flourish again.

Predators live in all ecosystems. They inhabit woods, mountains, deserts, suburbia, cities, the sky, the water, and, in the case of domestic cats and dogs, even our own living rooms. Most animal predators are mammals, but birds and reptiles are significant hunters of prey as well.

Mammals

The word **mammal** was first used by Linnaeus in 1758, derived from the Latin word *mamma*, for "breast" or "teat." Besides the ability to nurse their young, this class of animals also possesses hair, three bones in the middle ear, and a neocortex — the part of the brain involved in higher functions. Most mammals are also **placentals**, which feed their offspring during gestation through a placenta and give birth to live young. Just five mammal species are **monotremes**, or egg layers. Like birds, mammals are also warm-blooded or able to maintain a stable body temperature.

Mammals are highly diverse and adaptable, living in habitats from land to sea to air. They vary in size from the largest sea and land mammals — elephants and whales — to the tiniest bats less than 2 inches long. Rodents, shrews, bats, whales, dolphins, hoofed ungulates, and primates, including humans, are all mammals.

Birds

With some 10,000 species worldwide, birds are both friend and foe to farmers. They are an essential source of meat and eggs, and they can be pests to both fruit and grain crops. While many birds are predators, only a few species are a threat to livestock, poultry, or pets. In addition to the well-known birds of prey — eagles and hawks — owls, black vultures, crows, ravens, jays, and magpies can also prey on farm animals and pets. More rarely, seagulls will attack and kill livestock.

Reptiles

More closely related to birds than to mammals, reptiles truly represent an ancient group of animals. Reptiles are often described as **ectothermic**, or cold-blooded, which is not quite accurate, as their blood temperature does vary, but they are generally unable to generate sufficient body heat on their own. This low metabolism conserves energy and therefore requires less food or fuel. Reptiles exist on far less food than equivalently sized mammals do. Reptiles also have eyes adapted specifically to daylight, although some snakes have heat-sensing organs as well.

Most reptiles are carnivores. While crocodilians, snakes, and turtles are not major predators in an agricultural setting, in specific situations they are a threat to domestic livestock, poultry, and pets.

Ever Evolving

We are continually learning more about animal evolution and the relationships among species and subspecies. Genetic research can be especially revealing, but it will be tested and refined as the dynamic process of science continues.

Vulpes vulpes, the red fox

Taxonomy

Taxonomics is a biological method of naming, classifying, and organizing groups of organisms by shared characteristics and evolutionary relationships. At first, physical scientists identified members of a species by describing attributes and behaviors. Appearances can certainly deceive, however, and more recently DNA sequencing is revolutionizing the determination and classification of species. It is also important to note that common names like *lion* or *hawk* may or may not correspond to the actual species.

Family. A group of related genera with the same ancestors

Genus. A group of species with the same ancestors

Species. A group of animals that are capable of producing fertile offspring

Subspecies. A group within a species that shares a unique geographic area, physical characteristics, or natural history

Clade. A group of organisms that are closely related and share a common ancestor

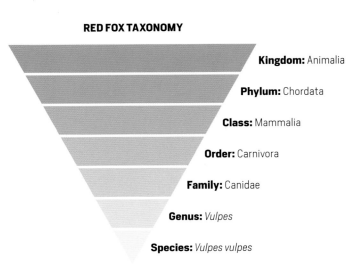

RED FOX TAXONOMY

Kingdom: Animalia

Phylum: Chordata

Class: Mammalia

Order: Carnivora

Family: Canidae

Genus: *Vulpes*

Species: *Vulpes vulpes*

Diagnosing Scat, Tracks, and Gait

By observing the "signs" of visiting animals, you can learn a great deal about what lives in your area, even before you acquire livestock. Record notes and take photographs for future reference and comparison.

Scat can vary in size and content for every individual, but it can reveal clues about the type of animal and its diet. It can be especially useful when combined with other signs. Both fresh scat and dry scat are potentially dangerous to humans. Handle either with great care, and do not inhale dry particles.

Track size varies with age and gender of the animal, regional or subspecies variations in size, the surface of the track impression, and the weather. In this book, track sizes measure the deeper minimum impression, not the more variable, broader surface outline. Tracks may also reveal tail drags.

Gait, or the way an animal moves, leaves a typical pattern of prints and can be classified into broad predator types of walking, trotting, waddling, and bounding. Stride impressions also vary individually and with terrain. Animals may walk or trot with a **direct-register** (rear foot lands in front print), **overstep** (rear foot lands beyond the front print), or **understep** (rear prints behind front prints). Canines often side-trot with an angled body, their hind prints to one side of their front prints. In a **straddle** trot, the hind tracks go to both sides of the front tracks. Strides are measured from the point one foot touches the ground to the next point of that same foot.

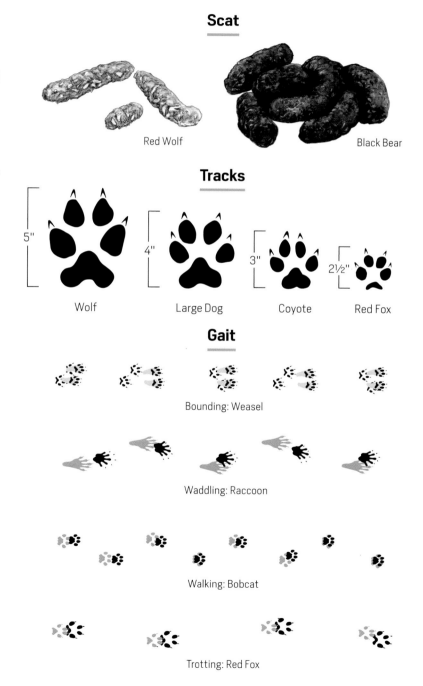

Scat

Red Wolf

Black Bear

Tracks

5" Wolf

4" Large Dog

3" Coyote

2½" Red Fox

Gait

Bounding: Weasel

Waddling: Raccoon

Walking: Bobcat

Trotting: Red Fox

CHAPTER 3

Canines

Canidae

The canine family is widely distributed around the world except for the Antarctic. In North America, this family includes wolves, coyotes, foxes, and domestic dogs.

Genus *Canis* originated between 4.5 and 9 million years ago. The timeline for the divergence of coyotes and wolves is in debate, with experts suggesting 1 million to 50,000 years ago. Although their genetic relationships are in flux, the living *Canis* species in North America include the coyote (*C. latrans*), the gray wolf (*C. lupus*), and the red and Eastern wolves (*C. rufus* and *C. lycaon*). Worldwide, other relatives include the jackals; the Abyssinian, Simien, or Ethiopian wolf; the dhole; the African golden wolf; and the domestic dogs and dingos.

OVERVIEW. The *Canis* species occupy habitats from tropics to tundra, forests, grasslands, deserts, and swamps. They share common characteristics and are medium to large in size. Their lithe bodies are adapted for chasing. They have large skulls, long muzzles, short erect ears, a naked nose, long legs, and a bushy tail. They are **digitigrade**, walking on the hairless pads of their toes (with 5 on front though the thumb is higher and does not make contact, 4 on rear) and with mainly nonretractable claws. **Carnassial** teeth (with an upper fourth premolar, lower first molar) are adapted for slicing skin, flesh, or tendons, and the molars allow the crushing of bones.

BEHAVIOR. The North American species share many common behaviors. They occupy a territory or home range, use scent signals and urine marks, and utilize dens for raising young but may often sleep outside. They are social, although individuals may live alone for part of their life. A mated pair is usually monogamous. Immature young need parental care, and many remain with their parents up to a year. A mated pair or the pack raises the young, with adult offspring continuing to help. Packs practice cooperative hunting and communicate with barks, growls, howls, yips, and whines.

Although the dog was a domesticated species, other *Canis* species were hunted for sport, for fur, or because they were predators that threatened livestock and poultry. Domestic dogs and gray wolves diverged in Eurasia as long as 40,000 years ago, although there was continued dog and wolf hybridizing. Domestic dogs are more closely related to the Eurasian wolves than to the gray wolf found in North America.

WOLVES

Modern wolves emerged in Eurasia 130,000 to 300,000 years ago and entered North America in multiple waves before the continents were separated about 10,000 years ago. Red and Eastern wolves may have been the first immigrants, entering North America before gray wolves. Many other wolf-like species are extinct today, such as the dire wolf. Following the extinction of the dire wolf, the gray wolf became the dominant species.

When the European colonists met the wolf, they knew it well. Widely hunted as predators of livestock in their homelands, wolves were extinct in England by the early 1500s, in Scotland in 1684, and in Ireland in 1786. They were gone from most of Western Europe by the early 20th century.

Wolves figured prominently in myth and fable around the world. In the New World, native peoples coexisted with the wolf and many tribes revered it. The explorer Meriwether Lewis (known from the Lewis and Clark expedition) described the wolves he encountered on the plains as "extremely gentle," perhaps because of this coexistence. Other explorers in the wilderness routinely noted that the wolves were particularly shy.

Wolf attacks on livestock increased after the European settlers hunted deer, bison, elk, and other hoofed animals and seriously depleted their populations. The colonists responded by offering bounties for the dead wolves. Today the list of subsequent wolf exterminations in North America reads like a death knell. By 1900, wolves were gone from the eastern states, except in the far northern Great Lakes region. We exterminated the Great Plains and Newfoundland wolves by 1926; the red wolf in the wild by 1930; gray wolves in Washington by 1940, in Colorado and Wyoming by 1943, in Texas by 1970, in Arizona by 1975, and in New Mexico by 1976. We caused the extinction of the Mexican wolf in the

▲ **Dire wolf.** This dominant predator, with a massive head and huge teeth, was found widely across North America. Existing alongside the American lion and the saber-toothed cat, the dire wolf became extinct with the other megafauna about 8,000 years ago.

The Europeans used words they were familiar with. *Wolf* is based on the Old English and Germanic *wulf*, *lobo* is Spanish, and *lupus* is Latin.

wild by the 1980s. We hunted several subspecies into complete extinction.

Humans poisoned, trapped, or shot the wolves. Pups were gassed or exploded with dynamite in their dens. Eventually we hunted wolves from the air. Today we have difficulty understanding how we allowed the very last members of a species to be relentlessly driven into extinction.

In the lower 48 states today, wolves are reduced to 5 to 10 percent of their historic range, with a total population of about 5,600 animals fragmented into separate populations. Transient wolves occasionally appear in other states, such as the lone Great Lakes wolf shot in Kentucky in 2013. Only in Alaska and Canada do wolves occupy much of their historic range.

SPECIES AND RANGE

Although debated, there is growing acceptance that 3 species of wolves exist in the United States: gray, red, and Eastern, although red and Eastern wolves may be **conspecific** (the same species). Surviving subspecies of the gray wolf include the Mackenzie Valley, Great Plains, Mexican, and Arctic wolves, and perhaps others in Canada. Experts note, however, that the taxonomy of wolves is complicated and under debate. Using genetic information and the increasing knowledge of behavior and physical characteristics, zoologists are reevaluating the species, subspecies, and historic ranges of wolves. Both scientific consensus and government classification are important for future conservation efforts.

Eastern and red wolves tended to live in the eastern and southeastern portions of the continent, respectively, while the gray wolves occupied more central and western areas — although there were certainly areas of overlap where the two species may have interbred. In the Great Lakes area, gray wolves may be the descendants of the Great Plains wolves, which are extinct elsewhere in the United States. A hybrid gray and Eastern wolf, sometimes called the boreal wolf or the Great Lakes boreal wolf, is still present today. The boreal wolf is larger than its

Current Wolf Range

- Gray wolf
- Eastern wolf
- Red wolf

Eastern relatives but displays the variety of coloration present in the gray wolf.

BEHAVIORS

Gray, Eastern, and red wolves share many common physical, social, and predatory behaviors. Built for hours of loping and bursts of speed, the wolf is also a good swimmer. It has acute senses of smell and hearing, and its sight is very responsive to movement. Wolves are intelligent and curious, quick learners, and problem solvers. Primarily nocturnal hunters, they rest during the day and become more active as dusk approaches.

Communicating both physically and vocally, wolves exhibit highly social behavior. They express dominance and submissiveness but also joy and playfulness.

LIFE CYCLE

The mated pair remain together for their lifetimes, often living and hunting with their extended family of pups and young adults 1 to 3 years old. The

▲ The Cherokee did not kill wolves; a weapon used to kill one was regarded as possessed, and they believed that the brothers of the dead wolf would seek revenge for the death.

female is in estrus only 5 to 7 days each year. If she conceives, about 63 days later she will occupy a den to give birth to her pups. When the pups reach about 6 weeks of age, she moves them to a protected **rendezvous** site where they play and practice their hunting skills while the adults hunt. Pups spend the next year with the pack, whose members help train them in social, hunting, and survival skills.

This family pack is a cooperative unit and may include unrelated members, generally an immature wolf or a replacement mate; however, most packs consist of parents, their juvenile pups, and possibly one or more older offspring. Packs are often as small as 6 to 8 family members but may range up to 15 to 20. Seasonal changes occur in pack size, becoming larger in the fall when the pups are large enough to help hunt. As winter progresses, older offspring reach sexual and physical maturity, leaving to find mates or their own territories. In some areas, the availability of large-sized prey also promotes exceptionally large pack sizes.

Wolves howl to locate pack members or potential mates, to reinforce territory against other wolves, to sound an alarm, and to signal a kill to their pack mates. Instead of low, smooth tones, they use a higher-pitched howl while pursuing prey, and they may bark as they close in for an attack. They also growl and whine.

HUMAN INTERACTION

Wolves are a keystone species (see page 8), widely regarded as essential and integral to the ecosystem. Population dynamics are complex and affected by various factors, but in areas without large predators, unchecked deer or elk populations can explode with disastrous effects. While wolf recovery is important to the environment, their presence in high densities can also be destructive to native ungulates and livestock.

Cattle and sheep ranchers continue to struggle with changing or inconsistent wildlife management regulations regarding wolves. In some areas, protests have been raised against reintroductions of wolves. By contrast, dedicated wolf advocates and organizations support such reintroductions. Emotions can run high on both sides of the issue, and on whether to keep wolves on the Endangered Species List or delist them when populations reach the target recovery levels. Wolves elicit stronger and more emotional viewpoints on both sides of the debate over preservation versus depredation control than do any other predator species.

PREDATION PATTERN

Nationwide, wolves are responsible for far fewer deaths of livestock than coyotes and free-roaming dogs are, and wolf kills are also less frequent than those occurring in Europe. In wolf country, however, their impact is greater, with calves the most frequent victims. In North America, cattle remain more vulnerable to attacks because they are frequently pastured during spring and summer in large or remote areas without range riders or other protective tools or supervision. Livestock guardian dogs (LGDs), which have proved themselves with sheep raisers, are not as commonly used by cattle ranchers.

Wolves take adult cattle, sheep, horses, and other livestock less often. Although wolves rarely take poultry, in Minnesota large numbers of pastured turkeys in a single field have been lost during wolf attacks — to a great extent because the birds panic, pile into corners, and die of suffocation. Wolves may also exhibit **surplus killing** of large numbers of natural prey or livestock.

Rabies is rare in the wolf population at present. Although wolves are closely related to domestic dogs, most wolf encounters are aggressive, as the wolf sees the dog both as a challenge to its territory and as prey. Small- to medium-sized and noisy dogs are the most likely to be attacked by wolves.

Fur trapping for wolf pelts is still an economically important activity, especially in Canada.

Alpha Wolf

Most biologists and animal behaviorists no longer use the terms *alpha male* and *alpha female* to describe the breeding pair of a wolf pack. Animal behaviorist Rudolph Schenkel first coined the terms in his studies of captive wolves in zoos, where unrelated wolves were forced to share a small living space. The idea of a linear or top-down dominance hierarchy captured public interest and was wrongly applied to domestic dogs as well.

We have since learned that the leaders of a wolf pack are a mated pair with lifelong bonds, unless parted by death. Pack members are their offspring or family of various ages, with very few exceptions such as a replacement mate or a young adult. Family relationships are cohesive and complex and often relate to age or individual temperaments. They are also dynamic, and pack membership changes over time as older offspring disperse. Either the male or female may assume leadership at different times and situations.

The preferred terms now are *breeding male*, *breeding female*, *breeding* or *mated pair*, or *parents*.

The male leader of a pack rebukes a young adult.

Gray Wolf
(Canis lupus)

Gray wolves are found around the Northern Hemisphere but have been reduced to about one-third of their natural range because of hunting by humans, loss and fragmentation of habitat, and lack of prey species. In North America, gray wolves continue to have natural prey in abundance in their current range, unlike much of Europe.

SPECIES AND RANGE

Eurasian gray wolves migrated to North America multiple times in various forms, and these may be the ancestors of different subspecies. The ancestors of the Mexican wolves were likely in the first wave, followed by the Great Plains and finally the Mackenzie Valley wolves. Physical differences have evolved between the gray wolves of Eurasia and North America. In general, North American wolves are more robust, with shorter legs, larger heads, shorter ears, broader muzzles, and denser coats.

The largest canid species in North America, gray wolves were historically found throughout Alaska, Canada, the western and central United States except for coastal and central California and the Southeast. Intensive predator control efforts at the state and federal levels eliminated most gray wolves from the lower 48 states by the 1930s. By the 1960s, only a small population of gray wolves remained in northern Minnesota. In Canada, wolves were no longer present in the eastern provinces and most of the southern areas of Quebec and Ontario.

Today, gray wolves now occupy 90 percent of their historic range in Canada, where the population is estimated at 60,000. They are not found in New Brunswick, Nova Scotia, and the southern areas of eastern Canada. The Alaskan population is estimated at 7,000 to 11,000 in 95 percent of its former range. In the lower 48 states, gray wolves are again present in portions of the Great Lakes states of Minnesota, Wisconsin, and Michigan; the northern Rocky Mountain states of Idaho, Montana, and Wyoming; and the eastern portions of Washington and Oregon. Transient wolves may be found elsewhere, and a breeding pair was confirmed in northernmost California in 2015. Gray wolves are usually found in more remote areas or parks and wilderness areas away from human habitation.

Surviving North American subspecies include the Great Plains, the Mackenzie Valley, the Arctic, and the Mexican wolves. In the past, scientists identified as many as 24 geographic or regional subspecies, although present-day research indicates that some of these were unwarranted. Some subspecies are extinct, such as the historical Kenai Peninsula, Newfoundland, Cascade Mountain, and Southern Rocky Mountain wolves. The status of the Pacific coastal wolves is still debated, including the British Columbian wolf, the Raincoast Island wolf (or *seawolf*), and the highly endangered Vancouver Island wolf. Taxonomists increasingly view the Eastern wolf as a separate species, perhaps conspecific with the red wolf, although this is debated.

Gray wolves rarely breed with coyotes in the wild, unlike the Eastern and red wolf species, but they do hybridize with Eastern wolves, and this may be the source of any coyote genetics in the gray. Further research will continue to reveal more information about genetic makeup of the species and subspecies, as well as their relationships and evolution.

DESCRIPTION

Size. The gray wolf is much more substantial in build than a coyote or a fox, with less pointed ears and muzzle and stronger jaws. A ruff on the sides of the head enhances the broad face. The eyes are golden yellow. With long legs, large feet, and a long stride, the gray wolf places the hind feet in the tracks of the front. It can run from 35 to 38 mph for short bursts and also has the endurance to chase down prey over distance. Loping or trotting at 5 mph, it can cover 30 to 50 miles per day.

In the western population, males usually weigh 85 to 115 pounds but can reach 130, with females weighing about 20 percent less. The heaviest male on record, at 175 pounds, was killed in Alaska in 1939. Males range from 5 to 6.5 feet long and females from 4.5 to 6 feet long; shoulder height is 26 to 38 inches. In the Great Lakes region, male gray wolves are 70 to 110 pounds and females are 50 to 85 pounds.

Coat and coloration. The long, dense, and fluffy coat is well adapted to cold weather, although gray wolves in warmer areas will have less dense and coarser hair. Longer, crest-like hair on the shoulders is 4 to 5 inches in length and covers the ears as well. As for color, a grizzled or mottled gray or gray-brown is most common, but black, nearly white or blond, and reddish brown are also seen. Pups may be different colors in a single litter.

26–38 inches at shoulder

4½–6½ feet from tip of nose to tip of tail

▲ Gray Wolf

Color Tints

In wolves, black appears to have entered the gene pool through crosses with dogs belonging to native peoples 10,000 to 15,000 years ago. Black is more common in North America than elsewhere in the world, at about 30 percent of a given population. In the Greater Yellowstone population, about half of the wolves are black.

HABITAT AND BEHAVIOR

Although gray wolves prefer cover for resting in times of cold or wet weather, they can thrive in a wide variety of habitats, from tundra to coastal areas, from forests to more open grasslands or deserts. The size of the pack's territory depends on the abundance of prey, terrain, climate, and density of the total wolf population. In the lower 48 states, territories can be less than 100 square miles, while in Alaska or Canada they may encompass 1,000 square miles. The pack marks its territory by howling, urinating, defecating, and leaving other scent marks against other wolves, coyotes, and dogs.

A mated pair or pack members travel through their territory searching for prey, using animal trails, frozen lakes, snowmobile paths, train tracks, or roads, following each other single-file in deep snow. When they locate larger prey, they stalk and then confront it. They try to isolate their prey from its herd, usually selecting animals that are old, young, or weak. If a prey animal stands its ground against the wolves, they try to force it to run, although some prey animals can, by refusing to run, successfully bluff wolves to abandon their attack. The wolves pursue fleeing prey, sometimes over distance, and it often escapes them. Small prey is pounced on. In a symbiotic relationship, ravens often follow gray wolves on the hunt, and wolves look for ravens circling over a carcass.

The average pack size is 6, but gray wolves can form packs of 20 or more members. Very large packs are more often found in Alaska or Canada than in the lower 48, although in 2000, the Druid pack in Yellowstone had a documented 37 members. Large packs of 30 or more sometimes occur when wolves are following prey migrations or hovering near large herds during birthing season.

Wolf packs are primarily composed of a mated pair and their offspring of various ages.

PREDATION PATTERN

Single gray wolves or pairs are capable of killing large prey on their own. Caribou, elk, mule deer, white-tailed deer, and moose are the primary prey of gray wolves, but in some areas they also hunt bison, wild sheep, mountain goats, and muskoxen. Smaller prey animals include beavers, snowshoe hares, marmots, and smaller predators such as coyotes, foxes, and weasels. Gray wolves will even hunt adult black bears and their cubs, wolverines, young mountain lions, and young grizzly or polar bear cubs. They will fight grizzly bears in defense of their pups but usually defer over a carcass. Black bears and mountain lions will defer to wolves over kills.

Gray wolves will kill cattle and sheep as well as other domestic livestock, pastured poultry such as turkeys in Minnesota, and dogs. Occasionally gray wolves will hunt rodents, voles, or birds; more rarely they eat fruit, berries, or other plants; they have been observed eating fish in Alaska. If fresh meat is not available, gray wolves will eat carrion, which may attract them to slaughter or stock burial sites. It is extremely unusual for wolves to scavenge human garbage in the United States or Canada. In severe winters, they will attack strange or weaker wolves and eat dead members of their pack.

After a kill, the breeding animals feed first. Gray wolves gorge themselves when food is available; they need an average of 2.5 to 7 pounds daily but will eat up to 12 pounds at one time. They also go without eating for days when necessary.

LIFE CYCLE

The mated pair breed in late winter or early spring, depending on location. In areas of abundant prey, after a severe winter, or following a large human cull of wolves, lone male wolves (called **casanovas**) may breed with older pack daughters, although they do not form a pair bond. This behavior has been observed in Yellowstone and Denali parks and among Arctic wolves. Incestuous breeding occurs in highly isolated populations. Lone wolves are believed to be 15 percent of the population and have a higher mortality than members of a pack.

The female gives birth in a den, safely located away from outlying areas of the territory and near a water source. Dens are usually repurposed animal burrows, the natural shelter of rocky areas, or covered depressions, rather than purposefully dug burrows. They may be reused for many years. Average litters are 4 to 6 pups, and gray wolves are known to adopt orphan pups. Sexually mature around age 2, offspring often remain with the pack until 2 or 3 years of age. When young adults leave to find mates or their own territories, they will sometimes travel for hundreds of miles.

Unfortunately, many pups don't survive their first year and less than half survive to adulthood. The average life span of a gray wolf is 7 years in the wild, while a few oldsters live to 12 or so. Starvation, disease, parasites, other wolves, and injuries from prey or other predators all contribute to wolf mortality. Vehicle accidents are a major killer in certain areas such as parks, causing up to 90 percent of deaths in Banff National Park. Hunting and trapping, illegal poaching, and targeted removal are also responsible for many wolf deaths.

LEGALITIES

Gray wolves are protected as an Endangered species in the lower 48 states, and each state handles gray wolf depredation complaints differently. In 2009, federal legislation provided for state-run compensation programs for verified wolf losses to livestock. The Defenders of Wildlife supports Wolf Coexistence Partnership funds to help ranchers implement the use of range riders, LGDs, and other nonlethal livestock protection tools.

Gray wolves are often managed under state control, and regulated hunting may be allowed; in other areas, they are a federally managed species. In Alaska, gray wolves are a regulated game species and furbearer trapping is allowed. Canada continues wolf predator control programs; provincial regulated game animal and furbearer trapping are allowed.

◄ Roman mythology describes the founding of Rome by the twins Romulus and Remus, who were fathered by the god Mars, suckled by a she-wolf, and saved by a shepherd.

SUBSPECIES OF GRAY WOLVES

Great Plains wolf
(*C. lupus nubilus*)

Also known as the loafer or buffalo wolf, the Great Plains wolf once roamed through much of the western United States, the central and northeastern portions of Canada, and southeastern Alaska. It is now extinct in the United States, other than a possible remnant population in the Great Lakes states. Great Plains wolves are also found from Manitoba throughout eastern Canada and north into Nunavut. Medium in size, they are usually light colored although black also appears.

Mackenzie Valley wolf
(*C. lupus occidentalis*)

Also known as the Alaskan, northwestern, Yukon, or northern timber wolf, this subspecies ranges throughout Alaska, across western Canada, and south into Washington, Oregon, Idaho, Montana, and Wyoming. Mackenzie Valley wolves are large, ranging from 100 to 135 pounds.

Mackenzie Valley wolves reentered the United States into Montana from Canada in 1979 and were breeding by 1986. Wolves dispersed to neighboring states as well, with at least one making it to Yellowstone. As part of the Northern Rocky Mountains Gray Wolf Recovery Plan, Mackenzie Valley wolves captured in central Alberta were relocated to the Great Yellowstone area and central Idaho wilderness as an experimental population in 1995. Fourteen wolves were released in both Idaho and Yellowstone. In 1996, additional wolves captured in northern British Columbia were translocated, with 18 sent to Yellowstone and 20 sent to Idaho.

The US Fish and Wildlife Service (FWS) currently classifies the gray wolf as Endangered at the species level, except in Minnesota, where it is listed as Threatened. The most recent effort by the FWS to delist the gray wolf in some regions and to allow states to classify the wolf as a game animal was halted by a federal court ruling in 2014. The legal status of wolves has frequently shifted at the federal and various state levels, reducing clarity in the management of wolves and confusing the public with changing regulations.

Arctic wolf
(*C. lupus arctos*)

Believed to be a subspecies of the gray wolf, the Arctic wolf makes up a substantial population on the isolated, far northern islands of the Canadian Arctic Archipelago. A very small population of about 50 wolves is also found in the coastal regions of Greenland. Called the grey wolf in Greenland and the white or polar wolf in Canada, the Arctic wolf is not viewed as threatened.

Weighing 70 to 155 pounds and measuring up to 5.9 feet in length, the Arctic wolf is well adapted to its extremely cold environment. Arctic wolves

The Arctic wolf is highly adapted to its challenging environment.

have a thick, waterproof, insulating coat, colored creamy white to gray; small and heavily furred ears; small furred paws; and the ability to build a thick layer of fat on the body. Most occupy territories of 1,000 square miles of tundra, although some follow the nomadic caribou herds. Litters are born in dens dug into the dirt or snow, or in rocky caves.

Arctic wolves prey mainly upon muskoxen and Arctic hares, whose fluctuating populations in turn greatly affect the wolves' numbers. To a lesser extent, they will hunt for caribou, lemmings, Arctic foxes, birds, and beetles. They will also scavenge on carrion or at human garbage sites, which can bring them into contact with humans.

Mexican wolf
(*C. lupus baileyi*)

The Mexican wolf is the most genetically distinct and divergent subspecies of the gray wolf. Research has shown that Mexican wolves possess genetic markers unique to North America, no doubt inherited from the earliest gray wolf migrants (none still living in the Old World). Popularly known as *lobo* (plural *lobos*), they once inhabited wooded mountainous areas in southern Arizona, New Mexico, Texas, and south into central Mexico. Two very similar subspecies found in the same areas, *C. lupus mollonesis* and *C. lupus monstrabilis*, were driven into extinction in 1935 and 1942, respectively. Some zoologists believe they all were essentially the same animal, while others believe the Mexican wolf is completely separate.

Today the Mexican wolf is the most critically endangered wolf in North America. Local wolf bounties were established in 1893, followed by the federal predator control program, which eventually hunted, trapped, or poisoned the Mexican wolf into extinction in the United States by the mid-1970s. Some experts believe the breeding populations were actually gone by 1942, although individuals still crossed from Mexico. The Mexican wolf was placed on the Federal Endangered Species List in 1976, the same year the last Mexican wolf was killed in Arizona. The last Mexican wolf in Mexico was captured in 1980, although it is possible that isolated wolves survived for a few more years.

Scientists established a joint recovery project with Mexico in 1982. Today 240 to 300 individuals exist in 53 captive breeding facilities (zoos and wolf parks) in the United States and Mexico, all belonging to the Mexican Wolf Species Survival Plan. A recovery area was established in central Arizona and New Mexico and a small portion of northwest Texas. Since the population is based on just 7 founder animals, scientists carefully exchange breeding wolves to facilitate genetic diversity. They select wolves for release from stock that is still represented in the captive population, acclimatizing and vaccinating them before reintroduction.

Beginning in March 1998, scientists released the first 11 individuals in the Blue Range Wolf Recovery Area, with additional releases through 2014. As of 2015, approximately 83 wolves survived in the Apache-Sitgreaves and Gila National Forests in Arizona and New Mexico. Some experts feel more captive-bred animals must be released into the wild or the population will suffer from inbreeding.

Since 2000, additional wolves were released into the White Mountain Apache Tribal Lands, following an agreement with the tribe, whose members maintain control over wolves on their land. No wolves are present in northwestern Texas. Other sightings in this area occur; however, most are probably hybrid wolfdogs or former pet wolves released into the wild. None have been confirmed as Mexican wolves. Mexico has also released captive-bred wolves into designated areas in their country.

DESCRIPTION

Size. The smallest of the gray wolves, the Mexican wolf ranges from 5 to 6 feet long, weighs 50 to 90 pounds, and stands 25 to 32 inches at the shoulder.

Coat and coloration. Mexican wolves are richly colored black, brown to cinnamon red, gray, and cream, with light underparts. Solid black or white wolves do not exist in the population.

HABITAT AND BEHAVIOR

Mexican wolves live in packs of 4 to 8 individuals, consisting of the adult mated pair and their offspring. Usually the parents are the only wolves that breed. Breeding occurs in February, with 4 to 6 pups born in April or early May. The Mexican wolf prefers forested mountain habitats for their available cover, water, and prey and is not found in desert scrub or grasslands. Territories may be up to several hundred square miles.

The pack hunts together, often chasing prey over distances, but this varies according to terrain and available prey. The main prey is elk, which makes up 74 percent of the wolves' diet. The wolves also hunt white-tailed deer, mule deer, javelin, rabbits, and other small mammals. They will scavenge carcasses and occasionally may kill livestock, especially young animals.

The major cause of death among the wild Mexican wolves is illegal killing, both intentional and when mistaken for a coyote or dog, followed by vehicle accidents and capture in privately placed leg-hold traps. A few have been lost to other wolves or mountain lions, and poor pup survival may also be a factor. No historical or recent cases have been recorded of Mexican wolves attacking humans.

LEGALITIES

The Mexican wolf is listed as a federally endangered subspecies in an experimental population. Brought together by the US Fish and Wildlife Service in 2011, the Mexican Wolf/Livestock Coexistence Council comprises ranchers, environmental groups, tribes, and county coalitions. Their mission is to support ranching, a sustainable wolf population, and a healthy landscape. The Mexican Wolf Interdiction Trust Fund provides compensation to ranchers for verified losses.

Red Wolf
(*Canis rufus*)

The red wolf is the most endangered canid in the world. Declared extinct in the wild in 1980, the red wolf exists only in very small captive and reintroduced populations. First observed and recorded in Florida in 1791, it was popularly called the red or small wolf, although black individuals were not unusual. John James Audubon assigned the first scientific classification for the red wolf as a subspecies of the gray wolf, describing it as smaller yet more fox-like in appearance.

In more recent years, the red wolf has been regarded as a separate species from the gray wolf, although not by all experts. In 2015, genetic researchers proposed that the red and Eastern wolves were the same species, distinct from the gray wolf, existing historically in the eastern areas of the continent not occupied by either the gray wolf or the coyote. The existing captive and reintroduced population of red wolf has probably drifted somewhat, however, because of its isolation and the small number of founders in this genetic pool.

Before the arrival of European settlers, the red wolf may have occupied a range bounded on the north by the Ohio River Valley and Pennsylvania, on the south by the Gulf Coast, west to central Texas and Missouri, and east to the Atlantic. Some models suggest they roamed even farther northward. They likely inhabited various terrains, although the current population seems to prefer wooded swamps and forested river bottomland. If the red and Eastern populations prove linked, the total joint range may have been much larger.

Like all wolves, the red wolf was relentlessly hunted as a predator. Three possible red wolf

◄ Historical residents of the southeastern United States, red wolves were described as smaller than their gray relatives.

subspecies were recognized in the past: the Texas red wolf (*C. rufus rufus*); the Florida black wolf (*C. rufus floridanus*), extinct since 1930; and Gregory's wolf (*C. rufus gregoryi*), extinct by 1980. The Texas red wolf was pushed into marginal areas, severely reduced in number, and in increasing contact with coyotes by the 1950s. At this time the red wolf was still surviving in parts of Arkansas, eastern Oklahoma and Texas, and Louisiana. Experts observed that although they behaved like wolves, they functioned in the ecosystem more like coyotes, preying on rabbits, rodents, and nutria.

RESCUE

In the 1950s and '60s, the zoologist Ronald M. Nowak observed that coyotes were beginning to dominate the Texas and Louisiana red wolf range and that more hybrid coyote–red wolf crosses were being seen. As the alarm about the red wolf's disappearance grew, the US Fish and Wildlife Service (FWS) decided that captive breeding was necessary to save the wolves, followed by the goal of reintroduction.

Following the establishment of the Endangered Species Act, the red wolf was among the first species listed. Beginning in 1974, the Red Wolf Species Survival Plan was designed to conserve the surviving population and reintroduce the red wolf species. Efforts began to capture the survivors living in marginal habitats in eastern Texas and western Louisiana. Every captured animal had a heartworm infestation, and many were in poor health. After trapping and examining 400 wolves, only a small number proved to be acceptable because of the extensive hybridizing with coyotes. Just 14 wolves became the founding population, and some of those were related to each other, leaving an effective genetic diversity based on 8 animals.

The breeding project has grown into some 200 red wolves held in more than 40 captive breeding facilities. Most wolves are at zoos, which include the red wolf in the Species Survival Plan of the American Zoological Association, and at a few wildlife centers or wolf parks. Additional propagation efforts took place on three coastal islands, although only the program on St. Vincent Island in Florida is still active. The recovery location was established in 5 counties in eastern North Carolina, encompassing 1.7 million acres, 60 percent of which is in private hands. The area includes the Alligator River National Wildlife Refuge and was expanded to 2 other wildlife refuges and a US Air Force bombing range.

The FWS released 4 red wolf pairs in 1987. Following a lengthy acclimatization, the released wolves were vaccinated and outfitted with radio collars or implanted radio transmitters in pups. From the beginning this was a heavily managed population; wolves were recaptured for medical treatment and relocation when they began to frequent human areas or dispersed beyond the recovery site. The project continued to capture wild-born offspring and radio-collar them as well. After coyotes became established in the area in the 1990s, efforts were made to limit hybridization, including the sterilization of coyotes and returning them as placeholders in their territories. This effort is now discontinued.

A second release of 37 red wolves in the Great Smoky Mountains National Park began in 1991. After 26 wolves died or wandered out of the park, the program was suspended and any surviving wolves were relocated out of the state. Biologists believe the wolves could not locate enough prey within the park, and when they left in search of food they were hit by vehicles, shot, or poisoned.

Although there has been some opposition to continuing the reintroduction project in North Carolina, the residents of the state generally support it. Nonetheless, the remarkable effort to preserve and restore the red wolf now seems threatened despite the success of the wolves at establishing ranges and reproducing. Although the FWS still regards the red wolf as a separate species, the agency has temporarily stopped all future releases of red wolves and is reexamining its program.

FWS agents are wrestling with difficult questions, including the proper taxonomy and historic range, the viability of the captive population, the need for additional sites (especially with ocean-level rise in the recovery area), the battle against hybridizing with coyotes, and continuing coexistence issues in the release area. In 5 years, the population has decreased from 110 wolves to an estimated 45 to 50 adults in 2015. The project has yielded much important information and serves as a valuable model for future carnivore reintroductions.

Red wolves are protected as an Endangered species in North Carolina and Tennessee. They are federally protected as a listed species in an experimental population.

DESCRIPTION

Size. The smallest North American wolf, falling between the coyote and the gray wolf in size, red wolves range from 45 to 80 pounds, stand about 26 inches tall, and are up to 5.5 feet in length. The head is wide, with a broad muzzle, and large, tall, pointed ears. Also distinctive is a lanky appearance, with long, slender legs and large feet. The long legs and large ears are the most distinctive features, helping to distinguish red wolves from both coyotes and gray wolves.

Coat and coloration. The short coat is usually gray, black, and cinnamon-buff, grizzled on the back with reddish ears, head, and legs.

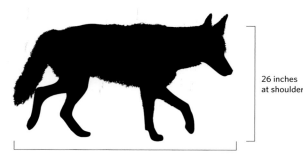

26 inches at shoulder

4½–5½ feet from tip of nose to tip of tail

▲ Red Wolf

HABITAT AND BEHAVIOR

Red wolves live in packs of 5 to 8. The breeding pair mate in February, and the pups are born from April to May in under- or aboveground dens, primarily in windrows next to agricultural fields. Most active during dusk and dawn, red wolves travel up to 20 miles per day in pursuit of prey. In North Carolina, white-tailed deer, raccoons, and rabbits make up 86 percent of their diet, varying by location and prey availability. Reclusive and shy, red wolves avoid contact with people, and no healthy animals have ever attacked humans in recorded history.

In the recovery area, red wolves have died due to vehicle accidents, disease or parasites, or territorial aggression by other red wolves. Since they resemble coyotes, many have been accidently shot by hunters. Hunting remains a threat to the success of the project. Red wolves have been removed from private land after becoming a nuisance.

▲ Despite passionate supporters, red wolf conservation has become controversial, as different factions debate the viability and integrity of the population.

Eastern Wolf

(*Canis lycaon*)

The Eastern wolf is also known as the Algonquin, the Eastern Canadian, the Eastern timber, or the deer wolf. Native peoples did not regard these wolves as a menace, but early colonists began to hunt them out of concern for their livestock. By the early 19th century, numbers were very small in New England.

SPECIES AND RANGE

Debate continues over the status of the Eastern wolf as a separate species or as a subspecies of the gray wolf. Currently the US Fish and Wildlife Service and the state departments of natural resources view the Eastern wolf as a distinct species. It has been confirmed that the Eastern wolf is not a hybrid of the gray wolf and the coyote; instead, evidence suggests that the Eastern wolf is historically the same as the endangered red wolf, **conspecifics** separated by loss of habitat and near extermination. This separation and the small captive breeding pool of the red wolf have no doubt set the two populations further apart. This changing categorization of both the Eastern and red wolves has ramifications for the protective status of both species.

Both the red and Eastern wolves and the coyote may have evolved from a common distant ancestor, separately from the gray wolf. This connection to coyotes may explain why Eastern wolves are more likely than gray wolves to mate with coyotes and produce hybrid pups. Historically, gray wolves and coyotes did not occupy the temperate forests in the eastern portions of the continent, which was the range of the Eastern wolf, although a gray and Eastern wolf hybrid may have existed in the past in overlapping areas of their ranges. This hybrid might be similar to the boreal wolf, a portion of the Great Lakes wolf population.

The Eastern wolves found in south-central Ontario and Quebec may be the "purest" representatives of the population. In Canada, authorities regard the Eastern or Algonquin wolf as a distinct and threatened species.

The highest density, fewer than 500 animals, is found within Ontario's Algonquin Provincial Park. A protected zone around the park now prevents the hunting and trapping of any wolves or coyotes. This measure has proved positive for the wolves, but the population needs greater protection and a recovery plan. The park is the site of long-term research utilizing radio collars and GPS to track the wolves.

DESCRIPTION

Size. More lightly built than a gray wolf, the Eastern wolf is a fleet animal that pursues white-tailed deer as its primary prey. Ranging from 62 to 77 pounds and standing 24 to 27 inches tall, the average female weighs 53 pounds and the male weighs 67 pounds.

Coat and coloration. Colored a grizzled black, gray, and white on the back, shoulders, and tail, the wolves have reddish brown shading on the muzzle, the back of the ears, and the lower legs. The flanks and chest are more reddish or buff. Wolves can be more reddish brown or gray but are not seen in pure white or black.

HABITAT AND BEHAVIOR

The small size of Eastern wolf packs, generally 5 to 7 animals, may be related to food resources. The breeding pair mate in February and raise their 4 to 7 pups with the help of the pack members. In late July or August, they take the pups to more open areas with some protective cover, such as meadows or bogs. Practicing their skills catching insects and small animals, the pups remain in these rendezvous areas while the adults hunt. Most offspring do not leave their pack until age 2, although some disperse at very young ages.

While Eastern wolves prey primarily on white-tailed deer, they also hunt moose and beaver. The wolves in Algonquin Provincial Park do travel outside its boundaries to hunt deer. Predators include humans, bears, and other wolves, although vehicles also kill a number of wolves. Since the 1970s, fearless or aggressive behavior toward humans has sometimes occurred in the park, primarily because of habituation.

Public Wolf Howls are a popular and educational event in Algonquin Provincial Park. The park wolves will answer human howls.

Where Wolves Thrive

There are areas on our continent where wolves seek to achieve the delicate balance of coexisting with other species (including humans) in their territory. Here are two examples.

Wolves of Yellowstone

Yellowstone, the country's first national park created in 1872, is nearly 3,500 square miles. Adjacent to the park, Grand Teton National Park occupies another 480 square miles. Together they are part of the Greater Yellowstone Ecosystem, one of the largest protected and nearly intact temperate ecosystems in the world. Its terrain is the site not only of volcanic activity but also of mountains, canyons, lakes, and rivers. Home to a wide variety of native wildlife, including our largest predators, the greater area is a vast natural laboratory. Surprisingly, for many years wildlife were not protected in the park and hunting was unregulated.

Inside the park, rangers killed the last gray wolves in 1926. In 1943, the last gray wolf was shot outside the park by a sheepherder. Wolves would continue to be seen in the area occasionally, but they were presumed to be transients, although some believed they were a remnant population of the original wolves.

continued on next page

continued from previous page

RECOVERY

In 1973, the US Fish and Wildlife Service (FWS) chose Greater Yellowstone as a recovery area for the gray wolf. Legally defined as an experimental population, the wolves were not subject to the same protections as an Endangered species. From 1995 to 1997, wildlife officials released a total of 41 wolves in the park itself — 31 from western Canada and 10 from Montana.

Some experts believed the introduced wolves were not representative of the historical subspecies, *Canis lupus irremotus*, the medium-sized Great Plains wolf still present in the Great Lakes. The actual relocated subspecies, the Mackenzie Valley wolf, is 30 percent larger and adapted to colder weather.

The wolf population quickly became established, increased, and dispersed to the surrounding areas under state control, which includes legal hunting. An estimated 400 to 450 wolves are now present in the Greater Yellowstone area, although residents believe there may be more. In 2013, about 100 wolves were counted in the park itself in about 10 packs.

SURVIVAL

Elk make up more than 90 percent of the Yellowstone wolves' winter diet. The pack also preys on bison, mule deer, coyote, moose, and smaller animals.

In the lower 48 states, the park is where people have the greatest chance of seeing wolves in the wild, and the wolves are indeed often visible to visitors. Slough Creek, Lamar Valley, Hayden Valley, and other areas are designated as wolf-viewing sites. Because of habituation to human presence, hazing or aversive conditioning is sometimes required so that wolves keep their distance from humans. Den sites are often closed to visitors.

As large as the parks are, they do not exist in a vacuum. They are surrounded by ranchland, livestock, development, towns, and people. Both prey and predators migrate outside the park, and their presence is often a contentious issue. Inside the park, the wolves are protected. Deaths occur mainly from aggression between wolves. Outside the park, wolves die in accidents, are legally hunted, or are poached.

Great Lakes Wolf Population

The Great Lakes region in the United States and Canada is the site of overlapping ranges of gray wolves, Eastern wolves, more recently arriving coyotes, and any hybrid animals that may have occurred between the 2 or 3 species. The hybrid gray and Eastern wolf is sometimes called the Great Lakes boreal wolf. Genetic research into these wolves continues to reveal more information, and it has been interpreted in different ways.

Historically, thousands of wolves lived in these states, preying on bison, elk, moose, and white-tailed deer. European settlers hunted both the predators and the prey as they timbered the land and turned it into farmland. Eventually the surviving wolves turned to livestock for food, and by the mid-19th century, state bounties were offered on wolves. By the mid-1960s, the wolf population had been exterminated in Wisconsin, reduced to about 20 wolves in Michigan (on Isle Royale), and cut to fewer than 500 in the northeastern corner of Minnesota.

Although the wolves were under some state protection earlier, in 1973 the Endangered Species Act placed what was then called the Eastern timber wolf (then viewed as a subspecies of the gray wolf) under federal protection in the Great Lakes region. The FWS developed a recovery plan, which is still treating both species as one group of wolves functioning as an apex predator. The wolf population in Minnesota dispersed on its own into Wisconsin in 1975, and then into Michigan. (The wolves are also found in the adjacent areas of Quebec, Ontario, and Manitoba.)

In the region, public support encourages the preservation of native animals and the wilderness areas, which are widely used for fishing, hunting, and recreation, while livestock owners are concerned about growing losses. Debate occurs over the reversal of delisting, again placing the wolves in Endangered or

Threatened status. The state departments of natural resources (DNRs) retain management responsibility for the wolves in any cases of depredation.

MICHIGAN

The Michigan DNR supports the presence of both the gray and the Eastern wolf in the state, as verified by two independent studies, and also a hybrid between the two. The DNR has found the wolves to be free from coyote genetics. In 2015, the population was estimated at a minimum of 658 animals, although the actual separate species numbers are unknown. The wolves are found in the Upper Peninsula, where they emigrated from Wisconsin. Both species prey on white-tailed deer and moose.

The separate population on Isle Royale National Park in Lake Superior, a 206-square-mile wilderness, originated in the late 1940s, when wolves crossed the frozen lake from Canada. Another male wolf emigrated from Canada in 1997. From a high of 50 wolves, the population averaged 25, although only 2 remained in 2016. The loss is primarily attributable to inbreeding, but the fluctuating moose population has also played a major factor in wolf survival.

WISCONSIN

The Wisconsin DNR reported a population of about 687 wolves in 2014. The state believes it can support a population of 700 to 1,000 wolves on both state and tribal lands. Depredations to livestock and hunting dogs do occur. White-tailed deer make up more than 50 percent of their diet, followed by beaver, snowshoe hare, mice, squirrels, muskrats, and other small mammals.

MINNESOTA

In 2015, the Minnesota DNR reported a stable population of approximately 2,200 wolves, in 470 packs. The wolves are found primarily in the northern forests and wooded lake areas but have begun occupying more of the central part of the state and dispersing into the south. They mainly take white-tailed deer and moose, followed by beaver and snowshoe hare, and they occasionally prey on small mammals and birds. As the wolves have begun to appear in agricultural areas, depredations to livestock and pastured poultry have increased. The state believes the wolf populations have recovered and that few suitable areas remain for the wolves to reoccupy.

▲ Great Lakes wolves live in relatively close proximity to humans, sharing forests, lakes, and recreational and agricultural land.

LGDs and Wolves

LGDs are one of best protections against wolf predation. LGDs must be mature and well bred, preferably from successful working parents, and they must receive correct training and experiences. They need to be supervised by shepherds or range riders and used in sufficient numbers for the situation.

The rule of thumb is that you need the same number of working LGDs as the wolf pack has members. Difficult terrain also requires more dogs. Failures of LGDs against wolves are linked to inadequate number of dogs, inadequate training and socialization to stock, use of night pastures near forested areas, and absence of supervision.

Wolf Watching

Follow wolf-watching guidelines at all times. Guided tours to observe wolves are also available in national parks.

- Park only at roadside pullouts.

- Do not approach or block a wolf's escape path.

- Do not entice wolves to approach you or offer food, even from your car.

- Use binoculars or telephoto lenses to get a close-up view.

- If a wolf approaches closer than 300 feet, get back in your car.

Hiking or Camping

- When hiking or camping, avoid den areas, which often smell of rotting meat scraps and urine. Avoid areas where ravens are circling overhead.

- Avoid campgrounds, picnic areas, or roadsides where wolves are habituated. Consult park authorities for current warnings.

- Supervise small children and leash your dogs while in parks and recreation areas with wolves.

Wolf Encounters

Despite the long history of fearsome tales, actual wolf attacks on humans are quite rare in North America. During the past 70 years in the wild, only 2 fatal attacks have occurred on this continent, both to lone humans in Saskatchewan and Alaska (2005 and 2010). Two people working with captive wolves were also killed. In contrast, pet wolves and wolfdogs were responsible for killing 9 children between 1986 and 1994, and attacking or biting others.

A small number of attacks have been recorded, resulting in nonfatal injuries or other nonphysical encounters in which the wolf acted in a threatening manner. Most of these have occurred in Alaska or Canada, with one in Wisconsin and another in Minnesota. Three of the physical attacks involved wolves later found to be rabid. The other attacks involved either habituated wolves or people engaged in recreation activities in wolf country.

Wolf Habituation

Habituation occurs when wild animals become so accustomed to humans they lose their fear of and reclusiveness from them. Since 1970, aggressive encounters with wolves have increased somewhat in parks and work sites in wilderness areas. This is due to both habituation and the recovery of protected populations.

Signs of increasing habituation include:

- Wolves on road, not moving for cars

- Wolves in residential areas, walking or resting

- Wolves resting and rendezvousing near humans

- Wolves approaching cars or people looking for handouts

Most problems with aggressive wolves are attributed to:

- Rabies, historically more significant in North America, and still a major cause in some areas of the world

- Habituation to people and/or food (see above)

- Provocation; self-defense; or cornered, trapped, or captive situations

- Interest in objects, such as backpacks, with testing to see if the object is prey

- Dogs with people

Wolves will defensively attack a dog, even on a leash, especially during May through July when raising pups. Baying or barking dogs may also attract wolves.

Reading a Wolf's Response

In the wild, when a wolf catches the smell, sound, or sight of a human, its natural reaction is shyness or fear. The usual prey of wolves walks on 4 legs, after all, not 2. The only animal wolves fear — the bear — will stand up on two legs like a human. A wolf may be curious about a human, but its next reaction is usually apprehension, not aggression.

A fearful wolf will place its tail between its legs and run away, perhaps stopping momentarily to look back at you. **Haze** or frighten the wolf as it leaves by yelling, waving your arms, or throwing objects. (See page 65 for more hazing tips.)

A defensive attack by a wolf is characterized by mock charges and barking. Since barking is driven by alarm, the wolf is likely protecting a carcass, pups in a den, or a pack rendezvous site. If a group of wolves tries to surround you, make a strong retreat without turning your back, if possible.

An offensive or predatory attack may have one of several warning signs: the wolf approaches at a walk, bounds around you, circles you, or bites at clothing or objects.

Dealing with Wolves
Homes and Yards

- Do not feed wolves or any other wildlife on or around your property.

- Hang suet feeders at least 7 feet above the ground.

- Dispose of garbage securely; clean outdoor food areas thoroughly.

- Prevent rodent infestations.

- Clear brush to eliminate cover and to improve visibility in and around your yard and children's play areas. Supervise children outside in wolf country.

- Feed pets indoors at night, and secure pet food containers.

- Keep dogs and cats indoors, well supervised, or in secure runs with a top. Walk dogs on a leash, and do not allow dogs to chase or harass wolves.

- Securely coop and fence backyard poultry.

- Install motion-sensor lights to offer some protection.

Livestock Husbandry

Wolves usually attack livestock in the summer when stock is out grazing and pups need feeding. Different terrain and stock situations dictate which methods, strategies, and tools are the best fit. Combinations always work better than one method, as well as changing tactics or adding scare devices to prevent habituation. Keep records to determine vulnerable times and pastures, as well as successful techniques.

- Monitor and protect animals during summer months, especially young stock.

- Remove stock from pasture early in fall.

- Remove unhealthy animals, carcasses, and carrion as soon as possible.

- Protect birthing areas and young animals; use night pens or other calving strategies.

- Use range riders, as human presence is very effective. Where permitted, haze wolves with air horns or nonlethal ammunition (paintballs, beanbags, and cracker shells).

- Avoid remote, heavily forested, or vulnerable pastures without shepherds or range riders and the protection of LGDs in sufficient pack size.

- Create **fladry** by suspending colored flags or fabric strips from a rope mounted along the top of a fence. Well-maintained, closely spaced fladry effectively deters wolves from crossing fence lines for up to 2 months. Turbo-electrified fladry remains effective 3 times longer than non-electrified fladry. For a full description on the use of fladry see page 254.

- Use other scare devices such as spotlights and automated or motion-activated lights and sound systems in birthing areas.

- Plan combinations such as LGDs, spotlights, and human presence.

Fencing

- Permanent fencing is always preferable but might not be feasible in very large grazing situations. Vulnerable livestock and poultry can be protected with temporary electric fencing.

- For pastured poultry, LGDs can patrol a fenced "no-man's-zone" between the perimeter fencing and the bird area; or the birds can be more securely penned or housed from dusk to dawn.

The Wolf as a Pet

Wolves are considered an exotic animal, and ownership either is prohibited or requires a permit in most states and provinces. Additional local regulations may also apply to keeping a wolf. Because of insurance issues, it is difficult to obtain veterinary care for wolves or wolfdogs and to insure their owners for liability. Rabies and other vaccines are not legally recognized for wolves or wolfdogs, which means that if a wolf bites someone it is subject to euthanasia for a rabies examination.

The inherent qualities of a wolf make life difficult, if not impossible, in a home. Sexual maturity brings behavioral changes, including a relentless curiosity and need to investigate, often in a destructive manner; a drive to roam, mark, and defend territory; and a predatory instinct toward all other animals, including pets and small children.

Most pet wolves spend their lives in secure pens, lacking all wolf social interaction and sufficient exercise. Unfortunately, many owners eventually euthanize their wolf or fruitlessly seek to place it in a zoo, wolf park, or sanctuary. Releasing a captive wolf is illegal and dangerous because of human habituation, potential disease transmission, and genetic contamination of protected breeding groups.

Wolfdogs

In the wild, mating between dogs and wolves is highly unlikely given the antipathy between the species, the social structure of the wolf, and the limited reproduction period. Biologists believe such breeding takes place primarily in situations where wolf populations are extremely small, limiting the wolf's choice of partner.

Historically, humans have utilized wolf genetics in some dog breeds, most recently in the Saarloos and Czechoslovakian wolfdog breeds. Wolfdog breeders generally use German shepherds, malamutes, and husky-type dogs to emphasize the wolf appearance in the crossbreed. Because it is difficult or illegal to actually own a wolf, most breeders use a wolfdog as the other parent, advertising the offspring as low-, mid-, or high-level wolf content. In reality, most dogs identified as wolfdogs turn out to be just crosses of wolfish-appearing dogs.

Owning or breeding wolfdogs is prohibited or regulated in many states and provinces. As with wolves themselves, the US Department of Agriculture has not approved any vaccines, including rabies, for wolfdog crosses. Wolfdog owners are often subject to insurance liability issues and local regulations. Claiming a dog as a wolfdog brings increased risks for its welfare. Shelters are also reluctant to place a wolfdog with adoptive owners, fearing future liability.

Without a doubt, wolfdogs possess unpredictable behavior patterns, depending on their specific inherited instincts and traits, and are not recommended as pets. Pet wolves and wolfdogs are responsible for far more fatal and serious attacks on humans than are wild wolves. These attacks are also bad for wolves in the wild, as they color public beliefs about the danger wolves pose to humans.

Wolf Attack: Defending Yourself

If a wolf approaches closer than 300 feet or acts aggressively:

- Do not run. Running triggers the chase and attack behaviors in wolves. Wolves can be deterred by aggressive or bluffing behavior by humans.

- Make yourself larger, raise your arms, and wave your arms, a walking stick, or an object.

- Pick up children or place them behind you.

- Do not pick up dogs because they are often targets for attack. Consider putting bells on the dog's collar when walking in wolf country to indicate to other animals that it is with people.

- Make noise, yell loudly, and throw objects — not to hit and harm the wolf but to scare him away.

- Back away. Do not turn your back on the wolf.

- Use bear spray if the wolf persists.

A wolf preparing to attack will stand with tail high and hackles raised or will snarl with ears flattened. Immediately react aggressively:

- Keep your eyes on the wolf.

- Step forward and make yourself bigger by waving your arms or a walking stick over your head.

- Throw objects at the wolf but don't take your gaze off the wolf to bend down to pick up an object.

- Yell loudly; use an air horn, other noisemakers, or firearms.

- Use bear spray.

- Stand your ground and continue aggressive actions unless there is a very close, safe retreat such as a building, car, or tree to climb. Continue to hit the wolf with your hands or objects if it attacks.

DAMAGE ID: Wolf

PREY ON

Young and small livestock, less commonly adult animals, pastured poultry, dogs

TIME OF DAY

Dusk through night

TRACK

Gray wolf: Front 3¾–5¾ inches long, 2⅞–5 inches wide; rear 3¾–5¼ inches long, 2⅝–4½ inches wide. Claws visible. A 2-month-old wolf pup will have a larger track than an adult coyote.

Red and Eastern wolves: Tracks are on the smaller end of the size range.

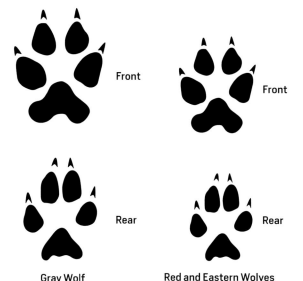

Front — Front

Rear — Rear

Gray Wolf — Red and Eastern Wolves

METHOD OF KILL

▸ Large animals attacked from rear, on sides, or on shoulders and brought down to the ground; tail or nose grabbed, bitten, and torn with large wounds; death occuring from blood loss or shock

▸ Smaller animals attacked on head, neck, throat, or back, with arteries or nerves severed or neck broken; may also be attacked in the same manner as larger animals

▸ Calves or laboring cows attacked in rectal areas or udder

▸ Carcass disemboweled and internal organs consumed first, including stomach lining. Rear parts fed on if wounded. Large pieces of flesh torn off and bolted down or taken to protected area for eating. Everything consumed except hide, large bones, skull, and stomach or rumen contents

▸ Carcass may be cached and returned to later, or may be abandoned to scavengers

▸ 70% of livestock kills are young calves up to 7 months, often when training pups.

▸ May surplus-kill sheep or other stock

GAIT

Stride when trotting about 22–39 inches or longer. May direct-register, overstep, or step slightly to the side.

SCAT

Similar size to large dog, cylindrical, tapered ends, about 6 inches long, 1 inch in diameter, black to brown, contains hair and bone fragments. Do not handle, as it may contain tapeworm eggs or other parasites.

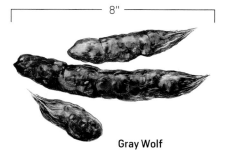

8"

5"

Gray Wolf — Red Wolf

COYOTE
(Canis latrans)

A central figure in Native American mythology and folklore, the coyote was often regarded as a shape shifter, a magician, or a creator. Old Man Coyote was the most widely known trickster in hundreds of tales of the Plains, Plateau, Southwest, and Californian peoples. Coyotes were occasionally kept as tamed pets in settlements and encampments. Farther south in Mesoamerica, the coyote carried an aura of military strength. Europeans adopted the Aztec or Nahuatl word *coyotl* in 1780, also using the names *brush wolf, prairie wolf, little wolf,* and *American jackal.*

Early Spanish reports first referred to coyotes as foxes or jackals. At the time of European settlement, coyotes were found primarily on the Canadian and American plains and also in the arid West into central Mexico. At various times, coyotes also ventured through the Midwest and east to the Appalachian Mountains, where explorers often described small, howling, and timid wolves. Meriwether Lewis described the "prairie wolf" in 1804, saying that it "barks like a large fierce dog." A year later, he wrote that the "small *woolf* or burrowing dog of the prairies . . . usually associate in bands of ten or twelve." The coyote was recognized as a distinct species in 1828.

Biologists consider the coyote to be a more primitive canid than the gray wolf, evolving in various forms in the Western Hemisphere and existing alongside megafauna like the dire wolf. A very large, wolf-sized, and more carnivorous coyote was in existence during the most recent Ice Age, but only

The word *coyote* is often pronounced with three syllables (*ky OH tee*) in the eastern and midwestern US states and with two (*KY oht*) in the western states and Canada.

a smaller coyote survived climate change, the resultant shift in prey species, and competition with the increasingly dominant gray wolves.

By 10,000 years ago the modern coyote had emerged. Nineteen geographical-based subspecies are recognized but probably not absolute given the coyote's fluid movements, adaptive success, and current evolving state. Following the near-complete removal of wolves from the lower 48 states, as well as the transformation of the land through deforestation and settlement, the coyote was able to take advantage of the new situation and spread widely.

In the late 19th century, coyotes began to move west off the Great Plains into southern California. Even earlier, they had begun moving farther north into Canada. Coyotes appeared in western Ontario in the early 1900s and in New York as early as the 1920s, and they were present in the larger Northeast by the late 1930s and '40s. They were observed in New Brunswick in the 1970s and Newfoundland in 1985. Similarly, coyotes were found in the southern states by the 1970s and '80s. Many urban areas in the East began reporting coyote issues in the 1990s. Today the coyote is found from Alaska and Canada throughout the lower 48 states and south into Panama.

Movement of Coyotes through North America

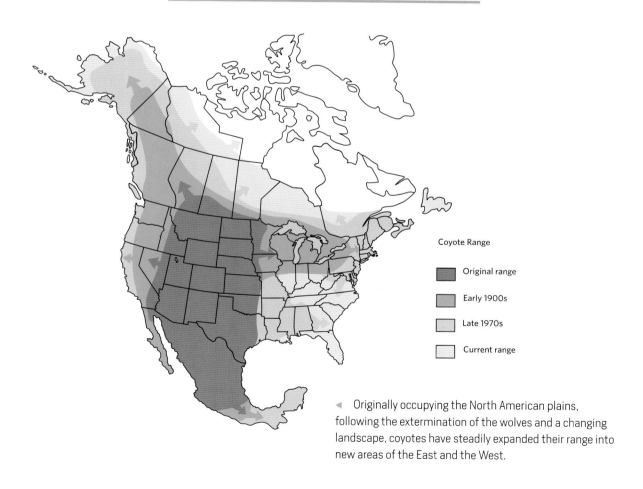

Coyote Range

▮ Original range

▮ Early 1900s

▮ Late 1970s

☐ Current range

◄ Originally occupying the North American plains, following the extermination of the wolves and a changing landscape, coyotes have steadily expanded their range into new areas of the East and the West.

Eastern Coyote

The most notable development of the coyote's expansion is the new subspecies known as the Eastern coyote (*C. latrans* var.), now genetically confirmed to be a hybrid of coyote, wolf, and domestic dog in varying percentages. The actual hybridization varies greatly, and the lack of stabilization or consistency means the Eastern coyote is not yet a separate new species, if it ever will be.

▲ Although it is popularly called the coywolf, biologists prefer the name Eastern coyote.

Coyote / Wolf / Dog Crossings

Earlier interbreedings may have occurred, such as with native American dogs, resulting in black coat color in both coyotes and wolves; however, the recent DNA evidence suggests that the primary crossings of coyote–wolf occurred about 100 years ago and of coyote–dog about 50 years ago. Both probably happened at the leading edge of the coyote migrations into new eastern territories.

After the near extirpation of the Eastern wolf, the few surviving wolves in the upper Great Lakes states, southeast Ontario, and the Maritimes may have mated with these immigrant coyotes out of desperation. Subsequently, when a few coyotes crossed from Canada into upper New York State about 50 years ago, they may have mated with dogs because of the same lack of available mates. In the southern states, the current coyote population may be the result of southern-moving Eastern coyotes, eastward-moving midwestern coyotes, and western coyotes from Texas and surrounding areas. The research suggests there may have been multiple **introgressions,** or backcrossing of genetics.

The illustration below shows how the different genetics show up in different species and regions in the East. Note that Western Great Plains wolf genetics are also present in some northeastern coyotes, suggesting a 4-way cross (Great Lakes/Ontario wolves, Great Plains wolves, coyotes, and dog). Unhybridized coyotes are found in the East, but genetic research has not revealed just coyote–wolf hybrids.

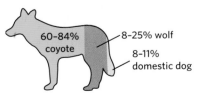

60–84% coyote — 8–25% wolf / 8–11% domestic dog

▲ Northeast

85% coyote — 2% wolf / 13% domestic dog

▲ Virginia

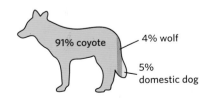

91% coyote — 4% wolf / 5% domestic dog

▲ Southern United States

Evidence does not support that this hybridization is still commonly occurring, although it is suspected in areas of red or Mexican wolves and in Newfoundland, where white (not albino) coyotes have been examined. Ongoing research will shed more light on the Eastern coyote.

Although popularly called the *coywolf*, this is not an accurate name and is rejected by most biologists. (See page 56.)

POPULATION

There may be no method to accurately estimate the coyote population today in North America; however, wildlife biologists believe it is increasing. Experts report that as many as 400,000 coyotes are killed annually with no appreciable effect on the overall population. Today the coyote is the most abundant livestock predator in North America and is responsible for about half of all cattle, sheep, and goat predator losses. Dealing with this situation has been a new and growing problem for poultry and livestock raisers in many areas of the country. Coyotes are also increasingly seen in urban areas, where encounters with people are now frequent.

DESCRIPTION

The coyote falls between the wolf and the fox in size, diet, behaviors, and social structure.

Coyotes have large, erect ears; yellow eyes with a round dark pupil; and a long, narrow muzzle. With a smaller skull and jaw than the wolf, the coyote has a weaker bite and is unable to grasp large prey the way a wolf can. Its teeth have larger chewing surfaces than a wolf's, indicating that the coyote also depends on vegetation for sustenance. It has slender legs, small feet, and a straight, bushy tail carried low or below level when moving.

Size. In general, the coyote is medium sized, ranging 42 to 50 inches in length from head to tail, standing 23 to 26 inches tall, and weighing 15 to 46 pounds. Significant differences occur among regional subspecies. California coyotes stand about

Coydogs

Despite widespread claims, wild coydogs are not common. While they are certainly possible from a biological and genetic standpoint, several major obstacles hinder both the creation and the survival of pups from a coyote and a dog. In addition, a natural antipathy exists between dogs and coyotes. Biologists believe that most coyote–dog hybridizing occurred when coyotes could not find another coyote mate, such as during the expansion into new territories. With the widespread coyote population in North America today, the likelihood of a coyote and a dog mating has actually decreased.

In the wild, it is a challenge for a hybrid pup to survive since it lacks the support of a father. The mother must provide all the food for her pups and leave them unprotected to do so. If a hybrid pup did survive, its reproductive cycle would not match that of other coyotes, due to a first-generation phase shift causing it to occur in November rather than January through March. Depending on its specific genetic makeup, a coydog may also lack the physical and instinctual traits for survival in the wild. Most supposed coydogs actually prove to be feral dogs or unusually colored coyotes.

Occasionally coydogs are bred intentionally in captivity; when that happens, the owner typically breeds a female coyote to a male dog. Various dog breeds may be used, resulting in a wide variety of appearances and behaviors. Coydogs are very challenging pets with unpredictable or mixed instinctual traits and without the wolf's social system behaviors. In many areas they are legally regarded as exotic pets and ownership is regulated. No rabies vaccine has been approved for coydogs.

18 inches tall, with males weighing 20 to 35 pounds and females 18 to 25 pounds. Eastern coyotes weigh 30 to 45 pounds, with some large males weighing 50 to 60 pounds and measuring up to 60 inches long. Adult female Eastern coyotes weigh 21 percent more than Western males. Larger from birth, with longer legs and larger skulls, Eastern coyotes are equipped to take larger game, such as the white-tailed deer, which are numerous in the East.

Coat and coloration. Coyotes are double-coated, with longer hair in northern areas of the country. They are commonly grizzled yellowish gray to grayish brown. The hair is banded in color with black-tipped guard hairs on the back and tail. The ears, muzzle, nape, flanks, legs, and paws are usually reddish. A black dorsal stripe runs down the back and across on the shoulder, with a black patch at the base and the tip of the tail. The coat is white, buff, or light gray under the throat and on the chest and belly. Coyotes can also appear dark brown to nearly black, blond or nearly white to reddish blond or rusty red. Eastern coyotes tend to be grizzled gray on the back, upper side, and neck, with a small number that are black or reddish blond. Visually it can be more difficult to distinguish between Eastern wolves and Eastern coyotes.

Senses. The coyote's eyesight is very sensitive to movement, and it has excellent hearing and a good sense of smell. The animals possess good endurance, running up to 40 mph for long distances, and they are capable of jumping an 8-foot fence.

Coyote Range
- Western coyote
- Eastern coyote

▲ Coyotes are found throughout much of Alaska and Canada except the arctic islands, the continental United States, and south to Panama. The Eastern coyote is found primarily in Ontario, Quebec, New Brunswick, Nova Scotia, Newfoundland, and Labrador; and in the United States, throughout New England, New York, New Jersey, and Pennsylvania.

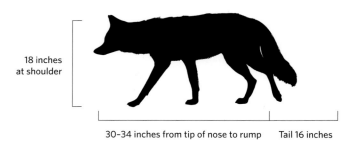

18 inches at shoulder

30-34 inches from tip of nose to rump · Tail 16 inches

▲ Western Coyote

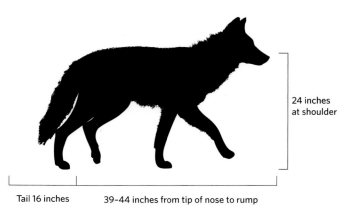

24 inches at shoulder

Tail 16 inches · 39-44 inches from tip of nose to rump

▲ Eastern Coyote

Vocalizations

Coyotes have a vocal repertoire that includes various barks, yips, yelps, growls, huffs, whines, and howls. The group will yip or howl when reunited, before they separate to hunt, or to declare their group territory. They will also howl in unison, which serves to create a social bond and to alert other coyotes to their territory. Lone howls are used to announce an individual's presence, express the energy of mating season, or call the group together. Howls are most often heard at night.

HABITAT AND BEHAVIOR

Historically, the coyote was found in open grasslands, brushlands, scrubby deserts, and alpine areas of nearby mountains. Today, although they inhabit areas from the tundra to the tropics — including dense forests, swamps, and higher elevations in the mountains — coyotes still prefer more open areas. Coyote density is highest on the plains and in the south-central United States. They are also found on agricultural land and suburban or urban environments from Central Park in New York City to Los Angeles. In more urban areas, coyotes will seek the cover of brush or wooded areas.

Coyote social structures are not completely understood and seem to vary with circumstances.

Transients usually live alone, but juveniles may form temporary groups, which are more aggressive or reckless than lone coyotes are with humans. Pairs may mate only for a season but more often have long-lasting or lifelong bonds. Mated pairs tend to hunt separately.

Coyotes are somewhat less likely to form packs than wolves are but far more likely than foxes. The pack is composed of an adult mated pair and up to 8 associates, which are often older daughters. Larger groups are associated with good food supplies, greater coyote density, and a lack of persecution, such as in national parks or other protected areas. The size and structure of the group or pack change over time.

Packs defend their territory, urinating and defecating to mark it. The home range size varies from ¼ square mile to 15 square miles depending on habitat, food availability, and the season. The animals use trails and dirt roads to travel their range.

Coyotes are most active at night and early morning, especially during hot weather or when near humans. They are seen more often during the day when raising pups, away from human areas, or when habituated to human presence or food sources. They rest in sheltered areas on the ground but don't usually use dens unless they are raising pups. Coyotes are less active in winter when there is deep snow or cold temperatures, conserving energy by resting.

Foraging and hunting. Opportunistic foragers and hunters, coyotes eat what their habitat and the season offer. Overall, 90 percent of their diet is mammals, mostly small animals from mice and voles to rabbits and ground squirrels. In California or western areas, their diet is 75 to 80 percent rodents and rabbits, plus insects, fruits, birds, and carrion. In other areas, coyotes primarily eat rodents, fruit, deer, and rabbits or lizards, snakes, and ground-nesting birds. Coyotes will consume grasses, fruits, berries, and garden crops such as cantaloupes, watermelons, and carrots, as well as utilizing garbage, pet food, and even cultivated grains.

When hunting small prey, coyotes slowly stalk and then pounce. They also chase small animals and dig for rodents. Coyotes have cooperative relationships with badgers, often digging near one another for prey. When coyotes capture prey in large amounts, they will cache the carcass.

Coyotes will hunt larger mammals, including deer, antelope, elk, and livestock. Usually preying upon the very young, they are the major predator of fawns in some areas. They will also pursue animals that are incapacitated or weakened by winter and animals that lag behind or travel on the periphery of a herd. Coyotes also work in groups, and they do so very persistently over time, either taking turns baiting and chasing their prey or driving it toward other pack members.

They tend to attack large animals at the hindquarters or flanks but may grab the head or neck and pull the animal down. Coyotes are far more likely to scavenge deer, however, than to kill them. In New York, a recent study confirmed that only 8 percent of consumed deer were actually killed by the coyotes; the rest were road- or winterkills.

Breeding. Breeding season occurs from January to March. Pair bonds are strong; coyotes court for 2 to 3 months before breeding and often mate for life. Female coyotes have only one heat period per year, which lasts for only 2 to 5 days and occurs in January through March. Male coyotes are fertile only during this time.

The pair will dig new dens or enlarge an abandoned burrow, preferring some slope to prevent flooding and to provide cover. Dens are also located in hollow trees, under logs or thickets, under ledges, or on steep banks. The entrance is usually 1 to 2 feet in diameter, with a tunnel passage leading to a larger chamber. The pair will reuse the den or make dens in the nearby area. Urban coyotes will den in storm drains, culverts, sites in golf courses or parks, and even abandoned buildings.

About 60 days later, 5 to 7 pups are born, on average, although very large litters of up to 19 are possible. Coyotes will produce larger litters in

▲ Coyote stalking prey. In areas without wolves, mountain lions, or bears, the coyote has become the dominant predator.

response to a drop in population. Both parents will care for the pups, along with other adults in a family group or pack. The coyotes will generally hunt alone during this time.

Coyotes are very protective of their pups in the area of their den. At around 6 weeks of age, the parents move them to another protected site. Male pups leave at 6 to 9 months of age, with both sexes mature at 12 months. Young females tend to remain with the group if possible, but do not breed.

Depending on the area, 50 to 70 percent of young coyotes do not survive to adulthood, often victims of birds of prey or other predators. If they survive, the coyote life span in the wild is 10 to 12 years. They are preyed upon by wolves — which do not coexist peacefully with coyotes — bears, and mountain lions. Coyotes themselves rarely confront bobcats, lynx, or foxes, but do kill them. In particular, foxes and dogs can be seen as territorial intruders.

Humans are the primary killer of adult coyotes, followed by disease and road accidents. In urban areas, vehicle accidents account for 40 to 70 percent of deaths. Research has repeatedly shown that increased coyote mortality results in higher reproduction rates, and that removing 70 percent of the entire population each year would be necessary to keep the coyote population at a reduced level.

There are no official sustainable or stable levels for coyotes, as there are for wolves and other Endangered animals like bears.

HUMAN INTERACTION

Important predators of many species (including rodents, rabbits, and white-tailed deer in the East), coyotes may also attack pets, poultry, waterfowl, domestic rabbits, and livestock. A coyote population may coexist without depredation for many years; however, killing the experienced adults will upset the stable situation, as will aggressive hunting, which drives younger coyotes into desperation.

Coyotes are more likely to take livestock in spring and summer due to both the presence of newborn stock out on pasture away from barns and close paddocks and the reality that they have pups to feed. Young coyotes in fall may also prey on stock. Coyotes will chew plastic irrigation pipes, leaving telltale areas of compressed or shredded plastic, and damage watermelons and other crops.

Urban coyotes are more likely than their rural cousins to habituate to people. They may be active in daylight, approaching people and pets and aggressively seeking handouts. Large numbers of urban coyotes are reported in southern California, generally in suburban areas adjacent to open recreation land. The Front Range in Colorado and the greater Chicago suburbs in Illinois also report large numbers of urban coyotes. Studies report that Chicago's coyotes eat rodents, fruit, deer, rabbits, birds, raccoons, and — far less often — garbage and domestic or feral cats. By contrast, in California, urban coyotes are definitely hunting cats; eating rodents, small animals, birds, and fruits; and scavenging for garbage or pet food.

LEGALITIES

Regulations vary by state or province: many areas have specific seasons for hunting coyotes as game or furbearer, some permit targeted removal of problem animals only, and others allow killing at any time without cause.

Coyote Encounters

Incredibly shy and reclusive, coyotes avoid human contact. In most cases, an encounter is a surprise to both sides. You are more likely to meet a coyote in urban, suburban, or recreation areas where they have become habituated to either food or people, than in a rural area.

A coyote minding its own business at night, or a coyote out during the day that flees from human contact, is not unusual or threatening. If you see a coyote in your yard or threatening your pets, you should institute prevention and hazing measures. Report fearless coyotes that approach people or act aggressively (growling, barking, nipping) to authorities.

If you are aware of coyotes in your area, tighten up your prevention techniques, exercise care when outside, and carefully supervise children and pets. Consistently haze coyotes around your home or nearby areas (see page 65).

On a Trail or around Your Home or Farm

- Remain alert to signs of coyote presence in the area. You are more likely to see a coyote from late fall to early spring, when food is more scarce.

- Do not confront a coyote or corner it, especially a coyote with food or in a den with pups.

- Walk in groups.

- Keep children close and supervised. Don't leave children alone when coyotes are known to be present. Children should be taught to recognize a coyote and what to do in an encounter. Because children often think the coyote is a dog, they need to be taught never to approach any strange dogs as well.

- Use caution around blind corners and covered areas. Jogging and bicycling are fast and quiet activities, lending themselves to more surprise encounters. Coyotes may chase a runner or bicyclist.

- Carry a whistle, an air horn, a walking stick or umbrella, a squirt gun, pepper spray, or bear spray.

- Report the encounter.

Dog Walking

- Keep in mind that dogs are attacked most often in winter.

- Walk your dog on a short leash next to you or slightly in front of you, even in parks or recreation land. Don't use extension leashes. Coyotes have attacked dogs on leashes. A dog off leash will chase a coyote, which is a very dangerous situation, as the coyote may be part of a group and not just a single animal.

- At night walk your dog only in lighted areas where other people are present and make noise. Avoid isolated or brush-covered areas or potential den sites.

- Avoid a strict schedule of walks, especially at night, since coyotes are very observant.

- Although small dogs are at most risk, larger dogs are sometimes attacked in the breeding season when coyotes are territorial.

- Don't permit large dogs to play with or act friendly to coyotes.

- Keep your dogs up-to-date on vaccinations.

COYOTE HABITUATION

There are escalating signs of coyote habituation and potential aggression in urban areas, where the animals are:

- Seen more often on streets and in yards at night.

- Killing more pets at night or approaching adults.

- Sighted on streets, in parks, or in yards from early morning to late afternoon.

- Chasing or killing pets in daytime.

- Attacking pets on leashes, and chasing joggers or bicyclists.

- Seen near children's play areas during the day, including parks and schools at midday.

- Aggressive to adults at midday.

COYOTE SIGNS

- Scat left on trails, rocks, logs, or conspicuous places near den

- Den sites with bones, fur, or feathers near outside entrance

- Calls, barking, yipping, or howling

SEASONAL BEHAVIOR

- **Late winter through spring.** Coyotes are defending den sites and pups.

- **Early spring into summer.** Coyotes are feeding pups and more active in daylight and are bolder in taking pets, poultry, or farm animals.

- **Fall.** Juvenile coyotes, singly or in groups, are roaming and looking for territory.

Dealing with Coyotes
Farm Buildings, Homes, and Yards

- Consult your local coyote regulations and work with your neighbors to reduce attractants and discourage coyotes. Report situations where coyotes are being fed. Haze coyotes whenever they appear. In western plague areas, handle and dispose of dead coyotes with care. Place the carcass in a plastic bag and spray it with insecticide.

- Do not feed coyotes or other wildlife. Use good bird-feeding practices, as spilled seed attracts rodents and squirrels.

- Control rodent populations.

- Don't feed feral cats, which are proven sources of food for urban coyotes.

- Do not leave pet food outside, and bring pets inside from dusk to dawn.

- Secure garbage and any animal feeds. Compost properly.

- Clear brush, woodpiles, and refuse near fences to improve sight lines in areas where you walk or children play.

- Remove ripe fruit and clear windfalls or rotting fruit from the ground.

- Don't plant foods coyotes eat in urban or suburban yards, including avocado, edible and ornamental berry plants or bushes, date palms, figs, guavas, lychees, and passion fruit.

- Note that water in the landscape is an attractant in arid areas.

- Bear in mind that motion-sensor lights and sounds may be useful but are not long-term protection if coyotes habituate to them.

- Use LGDs, an excellent deterrent in larger yards and properties.

Livestock Husbandry

The use of LGDs, good fencing, and husbandry practices in combination are most effective. On the range, good husbandry and use of LGDs, combined with targeted protection of birthing or young animals, are also effective.

LGDs must be mature; be well bred, preferably from successful working parents; and receive correct training and experiences. Failures of LGDs are linked to inadequate number of dogs, inadequate training, and absence of supervision. Difficult terrain also requires more dogs.

Guidelines for good husbandry include the following:

- Remove weak or sick animals.

- Keep records as to seasonal depredations and use of pastures.

- Protect birthing areas.

- Remove sick animals and carrion.

- Use range riders or herders.

- Remove cover where possible.

- Keep sheep or goats with larger animals such as cattle.

- Keep lights on in night pens and corrals.

- Employ motion-sensor lights and sound; though temporary, these are useful at high-risk times.

- Move parked vehicles and scarecrow devices frequently.

Fencing

Coyotes can go under, through, or over many fences. They can also climb or jump fences.

- Use exclusion fences for maximum protection areas — 5.5-foot-tall mesh, predator-proof fencing, with outward overhang with electric wire, buried apron (28 inches deep or L-shaped 6 inches deep and 12 inches outward), or barbed wire at ground level outside fence.

- Use electric fence — 13 strands of high-tensile wire with electric trip wires 6 to 8 inches above the fence and outside the bottom to discourage digging.

- Use electric enhancement of existing fence — includes electric trip wire outside the bottom and 8 to 10 inches above the top on a bracket extending outward.

- Use portable electric wire or mesh for temporary protection on the range, for birthing, or for night penning.

- Use coyote rollers — professional products or wire threaded through PVC pipe, installed on tops of solid fencing.

▲ Unless habituated to humans or food sources, coyotes prefer to remain undetected and will avoid contact.

Coyote Hazing Tips

Hazing or harassing discourages a coyote from settling into your neighborhood. Hazing in this circumstance is not cruel and does not harm the animal. It is essential to prevent a coyote from becoming habituated to humans, which will ultimately lead to the animal being destroyed. Coyote–hazing training videos and workshops are available.

When to haze a coyote

- If approached by a coyote in your yard, neighborhood, or in a park.
- If a coyote is beginning to frequent your yard or neighborhood.

Steps in hazing coyotes

- Very important: haze every time the coyote appears.
- Stand your ground where the coyote can see you clearly and maintain eye contact. Don't run away.
- Use a variety of tools — whistles, air horns, bells, pans, shaker cans, tennis balls, loud voices, umbrellas, super–soaker squirt guns with vinegar water, pepper spray or bear repellent, garden hoses, sprinklers, motion–activated lights or sirens.
- Throw objects toward the coyote but not directly at it.
- Make yourself big by raising your arms and waving sticks, tools, or jackets.
- Don't stop if the coyote freezes or runs a short way and hesitates.
- Act in an assertive and exaggerated manner, including yelling and jumping up and down.
- Continue if a second coyote appears.
- Continue until the coyote runs away.
- Different people should haze, since coyotes can become habituated to a person they recognize.

Precautions

- Don't haze a coyote that appears injured or sick. Call authorities.
- Don't haze from March through July, when coyotes may have pups in a den.
- Don't haze coyotes that are at a safe distance.
- Don't haze a coyote that appears unusually bold, not frightened, or aggressive.
- Don't haze large groups of coyotes.

Coyote Hazing Field Guide (printable brochure), projectcoyote.org.

Coyote Attacks

Attacks by coyotes have resulted in 2 fatalities in the past 35 years. One was a woman hiking alone, who was attacked by a group of coyotes in Cape Breton Highlands National Park in Nova Scotia, and the other was a 3-year-old child in California. From 1985 to 2006, there were 142 coyote attacks on humans throughout the United States and Canada, although most did not result in serious injuries. Roaming dogs are far more dangerous and are responsible for an average of 30 deaths and 4.5 million bites in the United States each year.

Most coyote attacks occur in the western United States, with California home to half of all attacks, followed by Arizona with 14 percent. Children are the victims of the most serious injuries, but children, men, and women are equally involved in attacks.

The majority of attacks occur either during recreational activities (camping, walking, or biking) or outside a home in a yard. Most occur January through April and at any time of day or night.

Most victims were able to scare off the coyote by yelling or throwing objects, or they were able to escape from the coyote. In one-third of the attacks, the coyote was being fed near the attack site, either intentionally or opportunistically.

The documented attacks fall into 6 categories:

- Offensive or predatory: the coyote chased or approached a person to bite (37 percent)
- Sleeping or immobile person received an investigative bite (22 percent)
- Rabid coyote (7 percent)
- Pet attacked with human nearby or attempting to rescue the pet (6 percent)
- Coyote defending itself, pups, or a den (4 percent)
- Unknown due to lack of details (24 percent)

Coyote Conflicts: A Research Perspective, http://urbancoyoteresearch.com/coyote-conflicts-research-perspective).

Coping with a Threatening or Aggressive Coyote

- Pick up and hold children and small pets, or place them immediately behind you.
- Do not run. Back away slowly.
- Maintain eye contact.

- Make yourself bigger, wave objects, open an umbrella, be loud, act aggressively.
- Get Mace, pepper spray, or bear spray ready and use when appropriate.

- Throw sticks or rocks toward the coyote if it approaches.
- If a coyote bites at a bag or other object you are carrying, drop or throw it.

DAMAGE ID: Coyote

Coyotes will scavenge on carcasses of animals killed by other predators or natural causes, so dead stock should be examined to determine cause of death.

PREY ON

Sheep, goats, young cattle, poultry, rabbits, cats, and small dogs. Will damage garden crops.

TIME OF DAY

Night and early morning most common, although day is possible

METHOD OF KILL

- Small animals attacked by a bite to the top of the head, neck, or back, leaving puncture wounds 1–1⅜ inches apart and tissue or bone damage

- Small carcasses carried off

- Large animals attacked on the flank or hindquarters, or grabbed by the head or neck and pulled down; suffocated by a bite to the throat behind the jaw and below the ear

- Bite marks left along the back

- More than one animal killed or damaged, but only one fed on

- Begins feeding at the flank or behind the rib cage, opens the abdomen, and eats the internal organs, including the rumen. Intestines may be removed and dragged away; dismembered body parts may be carried back to den.

- Clean, knife-like cuts left on carcass; bones picked clean

- Splintered and chewed bones left behind, along with blood, scattered skin, wool, bones, and tendons; larger bones, skeleton, and hide left intact

- Young coyotes make messy kills like those of a dog.

- An animal that escapes a coyote attack may have wounds to the neck, throat, flank, shoulder, or hindquarters. Its tail may be bitten off.

- 80% of calves killed are less than 1 month old.

- Calves attacked through the anus, abdomen, or nose; laboring mothers attacked through genitals or hindquarters

TRACK

Eastern coyote: Front 2⅝–3½ inches long, 1⅝–2⅞ inches wide; rear 2⅜–3¼ inches long, 1⅝–2⅜ inches wide.

Front

Western coyote: Front 2¼–3¼ inches long, 1½–2½ inches wide; rear 2⅛–3 inches long, 1⅛–2 inches wide.

Rear

NOTE: A coyote track is half the size of a wolf's. Dogs usually have a more splayed or open print, with toes pointing in different directions. Coyote tracks are more elongated or oval shaped, with a tighter impression than a dog's; claws may be less prominent. Coyotes direct-register at a trot far more than dogs. (See page 27 for comparison.)

GAIT

Trotting stride 15–26 inches, may extend to 41 inches. May overstep, direct-register, or trot slightly sideways.

SCAT

Twisted, rope-like, tips tapered, 3–4 inches long, ½–1 inch diameter. May be left on rocks, logs, and trails. Varies according to diet. May contain hair, feathers, bones, fruit, or seeds.

3–4"

FOXES

The Old English word *fox* is often used to describe someone cunning or sly. In many cultures around the world the fox is regarded as a magical, mischievous creature or a trickster. In some ways, unraveling the story of the fox in the New World is also tricky.

As both a native and an introduced animal, foxes are found throughout the world except in Antarctica. They are among the smallest members of the Canidae family. Two distinctly different genetic fox groups are found in the United States and Canada: *Vulpes*, the "true" foxes, and *Urocycon*, the more primitive gray fox found only in the Western Hemisphere. *Vulpes* includes the red, swift, kit, and Arctic foxes.

Both the red and gray foxes have triangular faces with erect, pointed ears; a long snout; partially retractable claws; and a bushy tail. Their slim legs are more suited to leaping and holding prey than to sprinting. Marking their territories with urine, foxes are usually solitary, but during breeding season a mated pair will share the responsibility for raising offspring.

Gray foxes are excellent climbers, able to clamber straight up tree trunks and leap from branch to branch. They may also raise kits in the cavities of trees, high above the ground.

LIFE CYCLE

Mating in late winter, the fox pair raise their pups or kits together and teach them to hunt at 4 months. The young leave their parents by late summer, although female offspring may remain in proximity. Vixens are very protective of kits, and a male that loses his mate will raise his kits alone.

Foxes often have short lives of only 4 to 5 years, although they may live to 10. The young are especially vulnerable, with 50 percent mortality due to birds of prey, bobcats, mountain lions, coyotes, wolves, bears, and even other foxes. Trapping and hunting remain the greatest threats to adult foxes.

PREDATION PATTERN

Often regarded as pests but also valued for their fur, foxes were trapped and hunted on foot or on horseback. They will prey on poultry, waterfowl, game birds, rabbits, and other small farm animals such as lambs, kids, or piglets. The red fox is most commonly responsible for poultry or livestock depredation, followed by the gray fox. The swift, kit, and Arctic foxes are typically not responsible for these problems, as they primarily eat rodents or live away from human areas. Many foxes readily adapt to a wide variety of habitats, including near humans in rural, suburban, and urban areas, where they will feed on rodents, garbage, or pet foods.

Both red and gray foxes are important predators of rodents, rabbits, and other small mammals. They may also prey on small adult cats or kittens, rabbits, guinea pigs, poultry, waterfowl, game birds, or small animals. More rarely, they may attack newborn animals such as lambs, kids, or piglets. Gray foxes may eat fish from ornamental or farm ponds. Both foxes will also eat eggs, grapes, fruit, and berries.

Fox furs are valuable: their pelts remain 45 percent of the total US fur harvest. Red foxes are more valuable for fur than grays, and large numbers are taken yearly.

LEGALITIES

Consult state or provincial regulations. Most states allow foxes to be hunted or trapped as game or furbearing animals, and many allow depredation removal.

FOXES AND HUMANS

The sport and culture of foxhunting on horseback originated in England and was adopted in Europe as well as the former British colonies. Colonists imported foxhounds to Maryland in 1650, and transplanted English foxes were released in various regions.

Foxes were widely farmed for fur in the United States, selected for desirable color mutations such as silver and platina. Although the animals are still trapped in North America, most fox fur is raised in Scandinavia.

In the Belyaev farm fox project in Russia, silver foxes were raised for 50 generations and selected for friendliness, resulting in several dramatic physical changes (mottled coat, raised tail, estrus occurring every 6 months instead of annually) and a reduced fear of humans. Captive wild foxes are not recommended as pets.

Red Fox
(*Vulpes vulpes*)

The most widespread and abundant of wild canids, red foxes are found throughout the Northern Hemisphere. Those in North America, however, have been isolated from the Eurasian red foxes for more than 400,000 years. The native red fox's historical range included the boreal forests of Canada, the West south to Wyoming, and the northeastern United States, but it was absent from much of the middle and southern regions. Lacking red foxes to hunt in many of the eastern areas, colonists imported English red foxes to several locations in the 18th century.

Although similar in appearance, the native red fox has been separated from its Eurasian relatives for more than 400,000 years. Like the coyote, the species has also expanded its range.

For a long time, scientists assumed that the red fox population found today throughout the country was descended from those imports. Recent genetic research, however, has not revealed any English red fox presence in the native population, which is descended from the true native North American red fox. It is assumed that most of the original imports perished or failed to breed with the native red fox.

In the 19th century, the native red fox found in the Northeast extended its range southward and the Canadian population moved into the Midwest. In the 20th century, the native red fox also expanded its territory higher into the subalpine and alpine meadows of western mountains. In some locations, the red fox population may also descend from escapees from fur farms or deliberate relocations for hunting purposes. Today, the red fox is found throughout most of Canada and the United States with the exception of the arid Southwest.

Because red foxes prefer open country with nearby cover, their range grew as timbering, farming, and settlement transformed the North American landscape. They are found in suburban and urban areas, but higher densities occur in areas of mixed farm and woodlands. Red foxes have also expanded into areas formerly occupied by wolves or coyotes, which they typically try to avoid, but they also find themselves interacting more with gray and Arctic foxes.

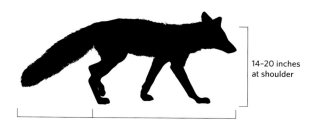

14–20 inches
at shoulder

Tail 12–21¾ inches   17–36 inches from tip of nose to rump

▲ Red Fox

▲ Red foxes are found across Canada except for the far northern islands and across the United States, including Alaska, except the arid Southwest. They are absent from areas of the western coast of both the United States and Canada.

DESCRIPTION

In addition to the American red fox found in eastern Canada and the United States, several subspecies occur that vary in size, physical characteristics, and coloration. These include the Labrador, Nova Scotia, British Columbia, Northern Plains, Cascade Mountain, Wasatch Mountain, Sierra Nevada, Northern Alaskan, Kodiak, and Kenai Peninsula foxes.

Size. Depending on the subspecies, the red can be the largest of the foxes. Red foxes are long-bodied, long-limbed, and lightly built with large pointed ears, a narrow muzzle, and slit-like pupils. The bushy tail is longer than half the body and reaches the ground when relaxed. The red has larger feet than the gray, and more prominent claws. They weigh from 6 to 31 pounds and stand 14 to 20 inches tall, 17 to 36 inches long. The tail adds 12 to 21¾ inches to the total length.

Coat and coloration. The coat is long, soft, and silky, but less dense in southern areas. Typically orangey-red, subspecies vary from pale yellow-red to deep red-brown. Red foxes have buff or grayish white underparts, distinctive black stockings, and a white tail tip. Different color phases of the red fox include albino, gray, silver, blackish brown, and the darkly colored spine and shoulder cross.

HABITAT AND BEHAVIOR

Highly adaptable, the red fox can be found from the tundra to the desert. Desirable habitat includes mixed hard- and softwoods, brush, grass, or cropland, since the fox prefers to hunt and live around the edges of both open and wooded areas, including suburban and urban environments. Red foxes tend to establish home ranges, although sizes vary considerably with available resources. Some foxes will wander without a range. The range of a male and one or two females may overlap, and they will defend their territory to a degree.

Hunting and foraging. Solitary and nocturnal hunters, red foxes are most active in the hours before dawn and in late evening, but they will also hunt on overcast or dark days or during the winter. They have good vision, which especially notices movement. While their hearing is excellent, their sense of smell is weaker than that of many dogs. They generally move at a trot.

These omnivorous animals eat rodents, rabbits, reptiles, birds, eggs, insects, and fruits, caching surplus food under dirt or leaves and protecting their prey from other predators. They stalk, crouch, and then pounce up to 16 feet to land on their prey, biting its neck and shaking it. They occasionally prey on newborns of larger animals and also eat carrion.

▲ In 1850, English red foxes were introduced to Australia for sport, where they have become a terribly invasive species and now number more than 7 million.

Red foxes use primary and emergency dens connected by paths to hunting areas and caches, often returning for many years to the same dens. They dig dens on the sides of slopes, ditches, and rock clefts, or they use abandoned dens of other animals.

LIFE CYCLE

Red foxes usually form a mated pair, although occasionally a male may mate with an additional female. The pair mate in late winter and, depending on climate, 4 to 6 pups or kits are born from March to May. Offspring are sexually mature at 10 months. When resources are sufficient, red foxes may form small family groups of 3 to 4 adults and their young, with older unmated female offspring often helping bring food to new pups. Pairs may occupy the same territory, of 150 to 1,500 acres, year-round. In the wild, the young are especially vulnerable to predators. Adult red foxes dominate all other fox species except the gray.

Gray Fox
(*Urocyon cinereoargenteus*)

A more ancient and primitive group of climbing foxes, the gray fox moved into North America about 3.6 million years ago. It is distantly related to the East Asian raccoon dog, which can also climb trees, and the African bat-eared fox.

Preferring to live in the dense cover of deciduous forests, wooded river areas, swamps, pinyon-cedar, and shrubland, gray foxes range from southern Canada throughout most of the United States except for areas in the Great Plains and the northwestern mountains. Historically, the gray was more common in the eastern United States than the red, although the latter has moved into much of the gray's range as the forests were cleared for farming. Gray foxes are still the dominant fox in the Pacific Northwest. With forest restoration, they are also found in the boreal forests of the northeastern United States and southeastern Canada, and near more urban areas.

▲ In Canada, gray foxes are found in the extreme southern portions of Manitoba, Ontario, and Quebec. Across the lower 48 states, they are found in most areas except the northern Great Plains and Rocky Mountains.

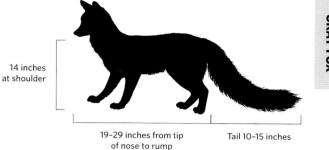

14 inches at shoulder

19–29 inches from tip of nose to rump

Tail 10–15 inches

▲ Gray Fox

DESCRIPTION

Although most often seen on the ground, the gray fox is noted for its ability to climb trees to escape predators and to search for food. Utilizing its strong, sharply hooked claws, a gray fox can climb 60 feet straight up the trunks of trees and then jump between branches.

Size. The 8 regional subspecies vary somewhat, but in general gray foxes weigh 8 to 15 pounds, stand 14 inches tall, and measure 19 to 29 inches long. The tail, measuring 10 to 15 inches, comprises about one-third of the total length. The gray fox has a strong neck, shorter legs than the red fox, and oval pupils.

Coat and coloration. The coat is grizzled salt and pepper on the back and yellowish red on the sides of the neck, the ears, in a line across the side, the legs, and the feet. The underparts are buff-white, and black stripes run from the eye to the corner of the head and down the muzzle to the mouth. The tail also has a black stripe, which ends in a black tip.

HABITAT AND BEHAVIOR

The gray fox is found more often in wooded or densely vegetated areas than is the red fox and has small home ranges of no more than 3 square miles. Solitary hunters for most of year, gray foxes eat fish, and they consume more insects and fruit than the other foxes, especially in summer. They are most active at dusk and night and sleep during the day in wood or brush piles, hollow trees, rocky areas, or near abandoned buildings.

Mated pairs use similar areas for raising their pups or kits, but some dens are located 30 feet high in hollow spaces of trees. Vixens bear 3 to 7 dark brown pups, born from March to May. Both parents care for their young, who are independent at 7 months. At 6 to 8 years in the wild, the life span of the gray fox averages slightly longer than that of the red fox.

OTHER FOXES

Arctic Fox
(Vulpes lagopus)

Marvelously adapted to its habitat, the Arctic fox is found throughout the northern pack ice, tundra, and boreal forest of Alaska and Canada, as well as Greenland and Siberia. Living a solitary life for part of the year, mated pairs will raise a large litter of pups together in large and complex tunnels and dens, sometimes with other related Arctic foxes. Hunting primarily at night for seabirds, eggs, lemmings, hares, and voles, the Arctic fox also eats berries and seaweeds and scavenges the kills of polar bears or wolves. Exceptional hearing allows Arctic foxes to find animals under the snow. They do not hibernate, and they must build up significant fat reserves before winter.

Size. Measuring 18 to 27 inches long and weighing 3 to 7.5 pounds, the Arctic fox has a compact and rounded body, a small muzzle and ears, and shortened legs.

Coat and coloration. An extremely dense, mutilayered, and plush coat; a very fluffy tail; and fur-covered pads on its feet protect the Arctic fox to −90°F. The Arctic fox has the best-insulated coat found in any mammal. There are two color populations: white in winter and grayish brown in summer; or blue-gray in winter and dark brown or bluish gray in summer.

Kit Fox *(Vulpes macrotis)*
Swift Fox *(Vulpes velox)*

The kit fox may be a subspecies of the closely related swift fox, and hybrids of the two can occur naturally. Their historical range was twice as large as it is today. The kit fox is found in arid territory between the Rocky Mountains and the Sierra Nevada, south into both Baja and central Mexico. Swift foxes are found on the grasslands east of the Rocky Mountains. They were reintroduced on the Canadian prairies in 1983.

Both the swift and the kit foxes are 12 inches long and weigh only 3 to 5 pounds. Pale reddish gray with a black tip on the tail, the kit fox has larger ears and a longer tail than the swift fox. Both hunt rabbits, rodents, ground squirrels, birds, lizards, and insects.

Island Gray Fox
(Urocyon littoralis)

There are two separate theories about the origins of the island gray fox. Sixteen thousand years ago, gray foxes may have floated or swum to the Channel Islands off the southern California coast. Alternatively, about 2,000 years ago native peoples may have brought them to the islands. They eventually inhabited 6 of the islands, and each of these fox populations is a separate, unique subspecies. Known collectively as the Channel Island or island gray fox, these little foxes were critically endangered due to golden eagle depredation in the 1990s, following the extermination of feral pigs, whose piglets served as prey for the eagles, as well as a lack of resistance to disease and parasites introduced by domestic dogs.

The process of insular dwarfism has reduced the size of the island gray foxes compared to their gray fox relatives. The foxes range from 2 to 6 pounds in weight and 22 to 31 inches in length, including the tail. The head and back are grizzled gray, with rusty red on the sides, white on the chest and underparts, and a black stripe on the top of the tail. The foxes subsist on mice, lizards, birds, insects, fruits, and berries.

Island gray foxes have made a remarkable and dramatic recovery since 2004, the result of collaborative efforts among the US Fish and Wildlife Service, Channel Island officials, and other stakeholders.

Dealing with Foxes

Rabies is a threat in the fox population, primarily in southeastern Canada and eastern and central states south through Texas and west into Arizona and New Mexico. Use great caution around any fox acting atypically, although foxes are too small to be considered a threat to humans and are not aggressive unless captured or handled roughly. In a fox encounter, the fox will attempt to flee. Gray foxes may also hide or climb trees to escape a threat.

Homes and Yards

- Remove all attractants to yard.

- Don't habituate fox to feeding or handouts.

- Provide backyard poultry and rabbits with good housing and pens.

- Prevent foxes from denning under porches, decks, or outbuildings.

- Use mild harassment (kitty litter, human-scented articles, lights, noise, or Mylar balloons near den entrance) to encourage parents to move a litter once the kits are old enough.

Livestock Husbandry

Spring is the most dangerous time for livestock, when foxes are raising pups and need more nourishment. Good husbandry practices include the following:

- Use shed lambing and more protected areas for birthing.

- Remove potential den sites or cover near animal areas. Foxes will raise pups right on farms, close to buildings, in pastures, or wherever they find cover.

- Use lights and noise only for temporary protection. Random or motion-sensitive devices may provide slightly longer-term protection.

- LGDs are excellent protection in fields and around pens.

Fencing

Foxes are good jumpers (up to 6 feet) and climbers.

- Exclusion fencing: Mesh fence openings should be 3 inches or less; include buried fence 1 to 2 feet deep or a 12-inch apron.

- Electric fence requires 3 strands, spaced (from the ground) at 6 inches, 12 inches, and 18 inches.

- Pens need a roof, mesh, or netting as protection, or an electric overhang.

DAMAGE ID: **Fox**

PREY ON

Rabbits, newborn livestock, poultry and waterfowl, eggs, cats. Garden crops. Will scavenge carcasses.

TIME OF DAY

Nocturnal, dusk; daytime when darkly overcast

METHOD OF KILL

Rabbit, newborn lamb, or kid:

▸ Killed by bite or multiple bites to throat, head, or back, and then being shaken

▸ Teeth marks: red fox canine teeth spaced ¹¹⁄₁₆–1 inch apart; gray fox, ½–¾ inch apart

▸ Usually one animal killed at a time; may be taken away and eaten elsewhere

▸ A hole torn open behind the ribs and the viscera eaten; nose, tongue, or head of small animals may be eaten

▸ A fox is not strong enough to crush skulls or large bones, or to hold down larger animals, which are often killed by multiple bites.

Poultry:

▸ Bird often missing, carried from the kill site to a den or cache site, leaving only a few drops of blood or feathers; bones found inside den or around entrance

▸ If carcass is present, breast and legs eaten first

▸ Toes curled (tendons pull toes up as meat is stripped from bones)

▸ Small bones bitten through

▸ Parts of the bird scattered nearby, with bones bitten through or partially buried

▸ Eggs opened in the nest and licked out, leaving shells

TRACK

Gray fox: Front 1⅜–1⅞ inches long, 1¼–1¾ inches wide; rear 1¼–1¾ inches long, 1–1½ inches wide. Claws not visible, no callus ridge.

Red fox: Front 1⅞–2⅞ inches long, 1⅜–2⅛ inches wide; rear 1⅝–2½ inches long, 1¼–1⅞ inches wide. Claws visible, ridge of callus visible across interdigital pad.

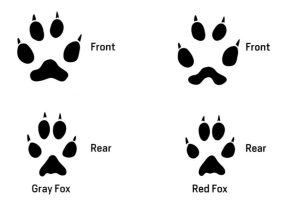

Front · Front · Rear · Rear · Gray Fox · Red Fox

GAIT

Gray fox: Trotting stride 9–18 inches. Often straddle-trots, rarely side-trots.

Red fox: Trotting stride 13–23 inches or more. Direct- and side-trot.

SCAT

Tapered tail; 2 inches long and ½ inch in diameter; may contain fur, feathers, berries, insects, plant fibers. Foxes also can transmit the *echinococcosis* tapeworm, potentially fatal to humans. Handle scat only with protection.

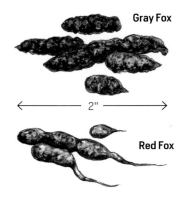

Gray Fox

←——— 2" ———→

Red Fox

CHAPTER 4

Cats

Felidae

Traditionally, the 40 species of cats or felines were divided into roaring and purring cats, regardless of differences in size.

The roaring *Panthera* group included the African lion, the leopard, the jaguar, and the tiger. The cats of the *Felis* group purred or made other sounds and included the small house cat (*Felis catus*), the lynx, the bobcat, the ocelot, the caracal, the jaguarundi, and the larger mountain lion. The cheetah and the clouded leopard stood alone in their own individual groups.

More recently, genetic studies have suggested that three separate lines of cats evolved from common ancestors: the small South American cats; the domestic *Felis catus* and other related cats; and the remaining large cats, including the lion, tiger, mountain lion, lynx, and bobcat.

What we do know is how similar the worldwide members of the cat family are to one another. Only the polar regions, Australasia, and Oceania have no native cats, although the domestic cat has been taken nearly everywhere. Cats resemble one another not only in appearance but also in their abilities and behaviors. Athletic and powerful, they are good climbers, and many are excellent swimmers. They have adapted to a wide variety of environments, and most display camouflage coloration. Cats are solitary, territorial, and polygamous, except for the sociable lions, which cooperatively raise their cubs.

The big cats are specialized carnivores, often apex predators near the top of the food chain and keystone species in their habitat. Capable of killing prey much larger than themselves, cats have long, conical canine teeth used for puncturing their prey and severing the spinal cord or killing by suffocation. All cat species except the cheetah have retractable claws.

Large eyes and excellent vision allow cats to hunt during the day or night. They both stalk and ambush prey, and they are fast sprinters but lack the endurance to chase prey over long distance. They have erect, movable ears, excellent hearing, and a good sense of smell. The rough tongue helps remove tissue from bones.

The cat's inborn hunting instincts — not a bloodthirsty nature — can result in **surplus killing** of more prey than it can eat. This instinct is not usually triggered in the wild, unlike the situation of enclosed captive domestic stock or poultry.

Many cats are perceived as dangerous animals and have been hunted extensively — out of fear, for sport, and often for their beautiful pelts. Habitat loss and human development remain great challenges to the survival of many feline species. The International Union for Conservation of Nature (IUCN) lists 29 of the 40 cat species as in decline, some critically endangered. In the United States, the jaguar, jaguarundi, ocelot, Florida panther, mountain lion, and Canada lynx are considered threatened.

MOUNTAIN LION
(*Puma concolor cougar*)

Mountain lions once roamed throughout the Western Hemisphere from the Yukon in northern Canada to the southern Andes Mountains in Chile. They were found in virtually every type of terrain, from high mountain forests to deserts to tropical wetlands. The mountain lion was present nearly everywhere in the lower 48 states.

Mountain lion, puma, cougar, or panther — the lord of the forest once ruled from coast to coast.

With this widespread presence, the big cat came to have more names than any other animal on earth — including cougar, puma, catamount, mountain lion, and panther. From Mexico southward it is still called by its Spanish names *leopardo* and *el leon*; indeed, Vespucci, Columbus, and Cabeza de Vaca all mentioned the "lions" they discovered in the New World.

For some time, European fur traders believed they were seeing only female lion pelts and that the larger males were hiding somewhere deep in the forests or mountains. Later it was widely believed that mountain lions were a light-colored African or Asian leopard or panther. In the Carolinas, Georgia, and Florida, it was often called *tyger*. *Catamount* or *cat of the mountains* was used from New England down the eastern coast. *Panther* and *painter* are still used in the southeastern states, although the mountain lion is definitely not a panther. In 1774, the famed naturalist Comte de Buffon first recorded the name *cougar*, probably evolved from Portuguese based on native Guarani, and *puma* itself was not used until 1858. Biologists finally settled on the name *Puma concolor*, or "one color."

Used as early as 1777, *puma* is a Spanish word and originated with a native Quechua word meaning "powerful." The Cherokee *klandagi* meant "lord of the forest," and the Chickasaw *keo-ishto* described the "cat of the gods."

Mountain Lion Ranges

- Western mountain lion
- Confirmed sightings
- Florida panther
- Confirmed sightings

◄ Mountain lions can be found in the western states from Texas north to southeastern Alaska, including the Rocky Mountains, the Cascades, and the Sierra Nevada. Breeding populations may be found in the western areas of Oklahoma, Nebraska, and North and South Dakota. Lone mountain lions have also been seen in Minnesota, Wisconsin, Michigan, Illinois, Indiana, Iowa, Missouri, Kentucky, Louisiana, and Arkansas, as well as reports from farther east, such as Connecticut. In Canada, mountain lions are found in the Yukon, British Columbia, Saskatchewan, and Alberta, with sightings in Ontario, Quebec, New Brunswick, and Nova Scotia.

While some sightings may be attributed to escaped pets or misidentification of bobcats, mountain lions are undoubtedly now reclaiming some of their lost habitat.

The "lord of the forest" reigned over the continent before the arrival of the Europeans. Many native peoples respected the big cat and viewed it as sacred, but the new colonists saw it primarily as a threat to livestock and game. Although they were familiar with wolves, mountain lions seemed exotic and dangerous. From the earliest days of settlement, local bounties were paid for killing mountain lions, and they were purposefully eradicated in many eastern states. As cattle and sheep moved onto the Great Plains, the federal government enacted predator control efforts based on hunting or trapping. Across the continent, development, habitat loss, and prey loss had a major impact on mountain lions; they need large home territories and the availability of their favored prey.

Once roaming the largest range of any land mammal in the Western Hemisphere, the mountain lion was gone from the eastern United States and Canada by the beginning of the 20th century, except for the very small population of Florida panthers. The Eastern mountain lion is officially regarded as extinct, although occasional sightings now occur in its old home grounds. Throughout the Western Hemisphere, the mountain lion now occupies about half of its former range. The species is protected in much of Central and South America.

Recent genetic research suggests the North American mountain lion is one subspecies, with five separate subspecies in South America. The cheetah, the jaguarundi, and the mountain lion are also

Florida Panther

The mountain lion in Florida has long been called *panther* and it reigns as the state animal. Although it was once considered a separate subspecies, *P. concolor coryi*, recent research placed the North American mountain lions together as one subspecies. This change did not affect the status of the Florida panther population, which is protected as an Endangered species. In 2013, the population of wild Florida panthers was estimated at 160, an increase from the 1970s, when only 20 to 30 panthers survived in the wild.

From its formerly widespread range in the southern states, the Florida panther now occupies only a fraction of that territory in 16 Florida counties, including the Everglades National Park, the Big Cypress National Preserve, and the Florida Panther National Wildlife Refuge. In this relatively small area, the population became fragmented and isolated, resulting in low genetic diversity. Eight female mountain lions from Texas were introduced into the Florida population in 1995. Although only 5 cats survived to produce offspring, the experiment was judged a success.

Threats to conservation include habitat loss due to development and traffic-related accidents. Florida panthers are also preyed upon by alligators and injured in territorial fights between the big cats.

genetically related. The mountain lion and jaguarundi may be descended from a common ancestor that first migrated into the Western Hemisphere some 8 million years ago. The modern North American subspecies itself may be descended from a small group that survived beginning about 10,000 years ago.

Today the mountain lion is reliably found in 12 western states, Alberta, and British Columbia, and its range is expanding. The population is estimated at about 30,000, although the actual numbers are in dispute. Mountain lions are hunted as game animals in some states and protected in others.

Solitary and reclusive, mountain lions are rarely observed even where their population is large, but their presence is becoming more evident in suburban and recreational areas and they remain a predator threat to livestock raisers. Humans and mountain lions are coming into increasing conflict, as development increases and subsequently fragments or isolates mountain lion territory. Lethal control has lessened in some areas, large-prey numbers have increased, and mountain lions are attempting to expand their range. Mountain lions may also become dangerously habituated to humans, necessitating their removal.

These big cats remind us of the power and beauty of nature. Balancing ecological needs and mountain lion management with human interactions, hunters, and livestock raisers will continue to be challenging.

DESCRIPTION

The mountain lion has excellent vision, hearing, and smell. It does not roar like the jaguar but can scream, purr, growl, and hiss. It can climb with great agility, jump as high as 15 feet, bound or leap as far as 25 feet, and sprint up to 50 mph. Although the mountain lion will cross a river, it generally avoids swimming.

Long and lean, females and males appear physically identical, making it very difficult to tell them apart. The strong legs have large paws equipped with retractable claws. The hind legs are longer than the front legs.

Size. Males are significantly heavier than females. Adult males typically weigh 130 to 150 pounds or more, and females weigh 85 to 120 pounds. Although more uncommon, larger mountain lions have been recorded at 190 to 210 pounds. The head and body are 3 to 5 feet in length, with a long tail of 24 to 40 inches. Adult males can have an overall length of 6 to 8 feet, including the tail. Standing 24 to 35 inches tall, mountain lions have a relatively small, rounded head with small, rounded ears.

Coat and coloration. Mountain lions are uniformly colored, either in shades of buff to tawny to red or silvery to bluish to slate gray. Different colors can occur in the same litter. An all-black color does not exist in the mountain lion population. The lips, chin, throat, and belly are lighter, with blackish tints on the muzzle, the rounded ears, and the tail tip.

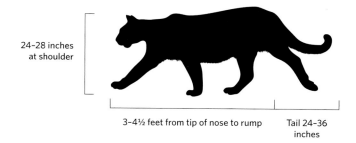

24-28 inches at shoulder

3-4½ feet from tip of nose to rump   Tail 24-36 inches

▲ Florida Panther   ▼ Mountain Lion

24-35 inches at shoulder

3-5 feet from tip of nose to rump   Tail 24-40 inches

HABITAT AND BEHAVIOR

Mountain lions are adapted to a range of habitats in North America, including open forests, brushy or rocky areas, canyons, wooded swamps, and grassland. In any habitat, mountain lions require sufficient cover and prey, but they will cross open, flat, and exposed spaces such as agricultural fields when necessary.

The territory or home range of a male mountain lion varies from 10 to 500 square miles, averaging 50 to 150 square miles. Size is related to prey availability and terrain. The male's territory may overlap the ranges of 2 or more females (which will be half the size of his), but not with those of other males. Female ranges may overlap one another as well, and females tend to raise their litters within the male's territory. Territories are marked with feces and urine-soaked debris or dirt piles, 4 to 6 inches high, often left near claw-marked trees.

Hunting and foraging. Both males and females without kits live solitary lives. They are primarily nocturnal, although most activity is at dawn and dusk. They can be active during the day in more remote areas or where they become habituated to human presence. Hunting mountain lions may travel many miles every night. They tend to cross dirt trails and roads rather than follow them, but they will use dry washes and trails to move between areas of their range or in search of new territory. During the day, mountain lions use many different resting sites rather than an established den, often moving on every night.

Silently stalking or ambushing its prey, the big cat will leap onto the animal's back and bite the base of its skull to break its neck. An adult makes a major kill about every 2 weeks, although scavenging birds and other predators are attracted to the kills, which may then force the cat to hunt again sooner. Mountain lions often drag the carcass to a cache site, covering it with debris. They return several times to feed and move the cache, usually at night. They often sleep near the carcass to guard it, although males may leave to patrol their territory.

Mountain lions clearly prefer deer and other large ungulates (elk, caribou, pronghorn, moose, wild sheep, and mountain goats) and can hunt animals much larger than themselves. Ungulates form from 70 to 90 percent of their diet, depending on location. Less often, mountain lions hunt smaller mammals and predators, including coyotes and bobcats. Mountain lions will also make use of the smallest mammals, birds, fish, or even insects and will also attack domestic equines, cattle, sheep, goats, pigs, rabbits, or poultry. They rarely scavenge carcasses. The Florida panther often hunts feral hogs and armadillos, while western coastal mountain lions will consume sea mammals.

LIFE CYCLE

Breeding is possible year-round, although in colder climates mountain lion kittens are more commonly born in spring and summer. Both sexes will yowl and wail at night to attract mates, and males will warn off other males. Mountain lions are probably polygamous breeders, and the female raises the litter alone. She uses caves, ledges, or dense cover for the initial period of intense care, but then moves the kittens frequently.

Litters average 2 or 3 kittens, which are born spotted and yellow-brown. Kitten spots slowly fade but are not fully gone until about 2.5 years, when they reach maturity. The kittens will remain with their mother for up to 2 years, which spaces litters 2 or 3 years apart. Young mountain lions do not breed until they have established a home range. Females may locate near their mother, but males may travel very long distances to find a range, occasionally far outside the established mountain lion ranges. Young adult littermates may travel and hunt together.

Mountain lion life span is 8 to 13 years in the wild. On average only 1 kitten survives from a litter, falling prey to other adult mountain lions, bears, or wolves. Adult males can have aggressive conflicts with each other, and both sexes become involved in deadly disputes over prey with wolves or bears.

Adults also lose their life to starvation, traffic accidents, ingestion of rodent poison, and hunting.

LEGALITIES

The Florida panther is protected as Endangered. Most western states permit mountain lion hunting but regulate it as with game animals; Texas, however, allows mountain lions to be hunted at any time. Mountain lions are protected in California and in central and eastern states, as well as in New Brunswick, Nova Scotia, and Ontario. Depredation permits may be issued in cases of human, pet, or livestock safety.

Beware the Hoax

Be aware that many false, inaccurate, or hoax stories and images of mountain lions circulate on the Internet. Forced perspective or altered images can make mountain lions appear to be incredibly large. Photos are often misidentified as to location as well.

HUMAN INTERACTION

Mountain lions are important predators of deer and other ungulates and a keystone species in the ecosystem. To a lesser extent, they also prey on livestock and pets and may attack humans. Although conservation programs have shifted the emphasis to coexistence and nonlethal control methods, sport hunting is allowed in some states, and depredation permits are issued for problem animals. While targeted depredation removes a problem animal, sport hunting can be counterproductive, especially when mature adult mountain lions are hunted as trophy animals. When a resident nonproblem adult is removed, a younger, inexperienced mountain lion moves into the area. In some cases, removal can be shown to increase predation on stock.

Mountain Lion Encounters

Human encounters with mountain lions are very rare, because the big cats are far more likely to avoid being seen. At times they may simply be curious or startled. Less commonly, humans stumble into a situation in which a mountain lion is hunting or has made a kill. Encounters are more likely where both the human and the mountain lion population is high. Coming upon a mountain lion face-to-face usually occurs in recreation areas or where housing is adjacent to mountain lion habitat that provides both cover and prey.

Mountain lion attacks have been increasing since the 1970s, with an average of 4 to 6 physical altercations per year in the United States. Although records vary, approximately 30 human deaths have been attributed to mountain lions since 1890, far fewer than the 20 to 30 deaths caused by dog attacks each year in the United States. The majority of victims are children. The most recently reported physical attacks have been in California, Washington, Colorado, and Vancouver Island, British Columbia.

While attacks are extremely uncommon, people who live, work, or enjoy recreation in mountain lion country should be sensibly prepared to deal appropriately with an encounter. Pay attention to local news regarding mountain lion sightings, attacks on pets, or habituated mountain lions. Signs of a habituated mountain lion include the animal being active during the day, showing no fear of humans, or approaching humans. Report all abnormal mountain lion behavior to local authorities and state departments of natural resources or the US Fish and Wildlife Service.

Attacks on humans are more common from dusk through dawn when the mountain lion is hunting or during late spring or summer when mountain lions are more desperate for food and/or young mountain lions are first on their own. Although younger mountain lions threaten people more often, adult mountain lions are responsible for more actual deaths. Children and smaller adults who are alone are viewed as more vulnerable. Children seem to attract a mountain lion's attention because of their size, quick movements, or high voices; therefore, they should be taught how to handle an encounter.

Recreation Guidelines in Mountain Lion Country

- Don't enter mountain lion territory from dusk through dawn.

- Look for active signs of mountain lions such as tracks, covered scat, or claw marks.

- Do not approach carcasses, which may be covered with leaves or other debris and may be attracting vultures.

- Remain in groups, not alone. Mountain lions are three times more likely to attack a solitary person.

- Make noise while walking, running, cross-country skiing, or biking in mountain lion areas. Slow down and be extra alert around blind corners.

- Do not allow children to run ahead or lag behind. The best position is right in front of an adult. A backpack is a safer method for carrying small children. The presence of children appears to increase the chances of an attack.

- Keep dogs on a leash. Although the presence of a dog can attract the attention of a mountain lion, it also seems to discourage attacks. Dogs also give advance warning of a nearby predator.

- Carry an air horn, pepper spray, a whistle, bells, walking sticks, and a strong fixed-blade knife. Individuals who must work in or frequent mountain lion areas may choose to carry a firearm if allowed. Firing several shots at a mountain lion that is attacking is effective, but a single shot is not.

- Store all food in airtight containers, especially when camping.

- Sleep in a tent or building, not out in the open.

Dealing with Mountain Lions

If you are in an area with mountain lion–human interactions, consider taking serious measures to discourage these animals from your yard and home, working with your neighbors if possible. Notice potential travel routes along washes, waterways, dirt roads, and trails. Be alert to environmental changes such as drought, fires, lack of prey, or increased habituation of a mountain lion near your home.

Homes and Yards

- Never leave children or pets unattended from dusk to dawn in your yard or when they walk to school or a bus stop.

- Eliminate attractants. In dry areas and drought periods, water also attracts mountain lions.

- Avoid any food sources or garbage that attracts rabbits, rodents, deer, and other wild animals that may in turn attract mountain lions.

- Don't feed or encourage feral cats, raccoons, squirrels, wild birds, and other potential prey.

- Don't use plants in your landscape that attract deer. Deer-proof landscaping is good protection against mountain lions.

- Feed pets inside or remove leftover food promptly.

- Eliminate the potential shelter of deep cover, spaces under decks, abandoned buildings, and refuse. Prevent access to other potential sites, such as large storm drains.

- Trim or prune bushes and trees, leaving several feet of bare trunks, especially near areas where children or pets play. Remove trees or branches that provide access to your yard over the fence.

- Remove cover on both sides of your perimeter fences, around gates and entryways, around your house, and in dark corners of your property.

This not only reduces cover for the mountain lion and its prey but also improves your sight line toward a predator approaching your yard. A 50-foot open clearance around your property is advised.

- Use bright outdoor lights, especially where you walk; motion-activated light and sound devices; and very loud music to repel mountain lions.

- Consider keeping alert dogs that will sound an alarm if a mountain lion is nearby.

- Keep in mind that one livestock guardian dog (LGD) will alert you to a mountain lion and will attempt to scare it away, but it will take more than one to successfully battle an aggressive mountain lion in defense of children or stock.

Livestock Husbandry

Mountain lion attacks on stock can increase when preferred prey animals are not available. Maintaining grazing in good condition supports healthy prey populations. Theoretically, more prey takes pressure off stock, but mountain lion populations may also increase. In areas of prime mountain lion habitat, livestock producers can have continual predation. Research has proved that predation complaints increase by 50 percent when adult mountain lions in the area are killed, allowing young, inexperienced mountain lions to replace them. While targeted removal of a problem animal may be necessary, aversive conditioning or hazing by experts can also be effective.

- Remove potential sources of ambush or travel. Remove nearby cover up to a quarter mile from the farmstead and clear brush near important birthing areas.

- Use more secure and open pastures for small or vulnerable stock.

- Keep one LGD to alert you, but keep at least a pair to repel or confront a mountain lion. Mountain lions will fight with coyotes and wolves, so dogs are at risk in an attack.

- Make use of temporary night penning on range, which allows for more effective protection by shepherds, range riders, lighting, and LGDs.

- Remove sick or injured stock and carcasses.

- Use intermittent or motion-controlled flashing lights and sounds, such as very loud music, to scare mountain lions away. Note, however, that they may become habituated.

- Use fully enclosed pens with a sturdy roof or barns as the only mountain-lion-proof housing.

Fencing

Fencing to prevent mountain lion entrance is necessarily formidable and expensive. Substantial woven wire, chain link, solid material, or electric wire fences must be at least 10 feet tall with a 65-degree overhang facing outward. These materials are usually practical only in small areas.

- Create overhangs and extensions of solid material, barbed wire, or electric wire.

- Place wooden fence posts inside fences to prevent climbing.

- Make sure poultry and rabbit coops and runs are securely built and well covered. Cover windows and other openings with securely attached ¼- to ½-inch-square hardware cloth or welded wire. Covers must support the weight of an adult mountain lion.

Mountain Lion Attack: Defending Yourself

Absolutely Do Not Do This

- Do not continue to approach a mountain lion — this will not intimidate it.

- Do not block potential escape routes or corner the mountain lion.

- Do not approach a feeding mountain lion, a mountain lion with kittens, or kittens who are alone even if they appear abandoned.

- Do not turn your back, engage in quick movements, or run. The mountain lion's instinct is to chase fleeing prey. Do begin to back away slowly.

- Do not crouch or bend down, which makes you appear more like a prey animal.

- Do not play dead. Even standing still makes you seem vulnerable.

Do this immediately when a mountain lion is sighted.

- Protect small children and prevent them from running or screaming. Pick them up if at all possible but without bending over. Put them on your shoulders to make you both look larger. Have smaller people stand behind larger people.

- Face the mountain lion and maintain intense eye contact.

- Make aggressive actions, which reduce the chance of an attack. Make continuous noise; shout loudly and firmly; use air horns, whistles, or bells. Shouting has been shown to be more effective than a gunshot. Pepper spray is effective aimed at the mountain lion's face. You may need to spray more than once.

- Appear larger by slowly opening or waving clothing or objects, such as opening and closing an umbrella or raising your arms over your head. Climb safely onto a higher object like a rock or log, if available.

- Throw rocks, branches, or objects at the mountain lion, without crouching down to pick them up.

- Back away slowly. If possible move toward open ground or a busier area with more people, vehicles, or homes. Seek a safe refuge but do not attempt to hide under something or to climb a tree.

- Remain alert if the mountain lion disappears. It may circle back behind you or follow you. Stay in open country.

Signs of aggression or an imminent attack by a mountain lion include:

- Crouching

- Intense staring

- Ears sharply forward or laid back

- Teeth bared

- Hissing or growling

- Tail twitching

- Hind feet kneading the ground

- Following your retreat

If a mountain lion attacks:

- Do not stop fighting.

- Hit the mountain lion with an object, your hands, or a fixed-blade knife — aiming for the eyes or face.

- Try to remain on your feet and facing the mountain lion. If knocked down, attempt to stand and protect your head and neck.

- Adult intervention and protection is absolutely necessary if a child is attacked.

- Never leave a victim alone. Other people in your group should attack and harass the mountain lion.

DAMAGE ID: Mountain Lion

In the rare cases when mountain lions take poultry, they will carry it off. Larger animals may be completely missing, either cached or consumed. Scavengers are attracted to the carcass and complicate accurate assessment of initial cause of death. Locating a carcass on large or rough grazing areas is difficult.

PREY ON

Horses, cattle (usually calves less than 200 pounds and 1 year of age), sheep, goats, and other small stock. Poultry predation is rare.

TIME OF DAY

Nocturnal but most active dawn and dusk; daytime where habituated to humans or in remote locations

METHOD OF KILL

- Back of skull or neck bitten, revealing puncture marks; kill with a bite from above, breaking the neck and spine or biting at the head; occasionally they will bite at the throat like a coyote. Bite marks from upper canines will be 1½–2¼ inches apart, generally larger than a coyote; lower teeth are ⅜–½ inch smaller. Bites are clean, not ragged, and cause deep wounds in this area.

- Claw marks on neck, face, shoulders, sides, or belly on larger animals

- Drag marks present, usually in a straight line for at least 10 yards up to 350 yards; carcass covered under brush, dirt, debris, or snow; rumen or organs often covered separately; scratch marks up to 3 feet in length. The mountain lion will return to feed again for several nights, often resting nearby, moving carcass and re-covering it after each meal until remains are left uncovered and abandoned.

- Hair and viscera removed from carcass before heart, liver, and lungs eaten; stomach untouched; large leg bones and ribs may be crushed; sometimes front quarters and neck or hindquarters fed on first

TRACK

Front, 2¾–3⅞ inches long, 2⅞–4⅞ inches wide; rear, 3–4⅛ inches long, 2½–4⅞ inches wide. Distinctive 3-lobed heel pad 2–3 inches in width. Mountain lions move lightly and may not leave clear tracks on dry or hard ground. No claw marks. Teardrop-shaped toes. Can identify left or right prints because they slant like a human footprint.

Front

Rear

GAIT

Primarily walk, with stride length up to 32 inches. Will overstep or direct-register (step in the print or overlap it) the front track at a walk. The stride at a gallop can be 3–10 feet, with leaps up to 25 feet.

OTHER SIGN

Claw marks on trees, 4–8 feet above ground. Long, deep, parallel scratches without much bark removed, unlike with a bear.

SCAT

Blunt ends, 4–6 inches long and about 1¼ inches in diameter (size may indicate the relative size of the mountain lion). Usually covered, may contain cut pieces of bone and hair.

Mountain Lion

6"

Florida Panther

JAGUAR
(Panthera onca)

Often called *el tigre* in Mexico, a name bestowed by the Spanish, the jaguar is the largest cat in the Americas and the third-largest cat in the world. It is an important symbol of power in the Aztec and Mayan cultures, and jaguar images are found on the artifacts of North American native peoples as well.

In 1544, John Cabot recorded that jaguars were present in Ohio and Pennsylvania. In 1799, Thomas Jefferson noted that a jaguar weighed 200 pounds. Jaguars were occasionally spotted as far away as southern California, Colorado, and North Carolina. In the early 20th century, jaguars were still present in Arizona, California, New Mexico, and Texas.

Widely hunted for pelts, jaguars were rapidly exterminated in many areas of the Americas. A population of 15,000 jaguars may still exist in the wild, although numbers are still declining rapidly due to loss of habitat and hunting. International protection was achieved in 1973. The last female American jaguar was shot in 1963, followed 2 years later by the last males. In northwestern Mexico, the jaguar population numbers about 80 to 120 animals, and since the 1990s, individual jaguars have been spotted every few years in Arizona, New Mexico, and Texas.

DESCRIPTION

The jaguar shares a distant common Asian ancestor with lions, tigers, and leopards. Built for power more than speed, it has a larger head and broader jaw than the mountain lion, along with a more powerfully muscled body; shorter, sturdy legs; and a shorter tail.

◄ Before the arrival of the European settlers, jaguars roamed the continent from the Middle Atlantic to the northern Pacific coast.

Size. Jaguars weigh 100 to 260 pounds and range from 3.5 to 6 feet long, excluding the tail, although individual animals can be longer. Females are slightly smaller than males, and size variations occur in different regions. The largest jaguars are found in the Brazil Pantanal, with smaller jaguars seen in more open habitat.

Coat and coloration. With a reddish buff to light gold base color, the black rosettes or rose-like spots provide camouflage on the upper body, with solid black spots on the head and neck, and off-white underparts. Black jaguars, which form about 6 percent of the population, actually have a black base color with black spots.

HABITAT AND BEHAVIOR

Although the jaguar more closely resembles a leopard, its behavior and habitat are closer to the tiger's. Dense tropical forests are preferable, but jaguars will also use swamps, scrubland, thickets, and coastal forests that are near rivers, streams, or still waters. Jaguars are not common in arid or higher mountainous areas. The protected jaguar area in the United States is mainly oak, pinyon pine, juniper woodlands, wooded riverways, and mesquite thickets.

Jaguars maintain a home range that varies from 4 to 65 square miles, with females occupying smaller territories within the much larger home range of a male. The jaguar roars deeply like a lion to identify its territory, and it also communicates through grunting.

Hunting and foraging. Nocturnal hunters, jaguars are most active during dawn and dusk, resting at midday and seeking relief from heat in caves, under large rocks, in or near water, or in deep vegetation. Pouncing from hidden cover or high in a tree, they can take medium to large mammals and reptiles, in addition to birds, fish, and opportunistically smaller prey. If they take prey at the water's edge, jaguars will follow it into the water, as they are excellent swimmers. Jaguars will also fish from shore, grabbing fish with their claws. They can easily drag prey 3 to 4 times their weight for distances up to a

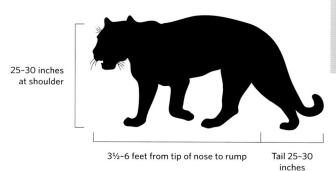

25–30 inches at shoulder

3½–6 feet from tip of nose to rump

Tail 25–30 inches

▲ Jaguar

half mile to reach a more secluded area. After hiding or burying the carcass, they will return to it to feed.

LIFE CYCLE

Female jaguars mature at 1 to 2 years of age, and males at 3 or 4. Jaguars are solitary animals except during estrus, which may occur year-round. Two cubs are most common, usually born in wet seasons when prey is more plentiful. They remain with their mother for about 2 years as they learn to hunt. The jaguar's life span in the wild is 11 to 12 years, and longer in captivity. Humans are the primary threat to jaguars through illegal poaching, trophy hunting, or killing by ranchers or farmers.

HUMAN INTERACTION

Essentially secretive animals, jaguars may take cattle, horses, or other livestock. They are not known to stalk humans, although attacks have occurred with habitat encroachment and loss of prey due to human hunting.

Now occupying less than half of their original range, jaguars are found primarily in 18 countries of Central and South America, from Mexico south to northern Argentina and Paraguay, northeastern Brazil, and the Amazon basin. The Brazilian Pantanal is home to the greatest density of jaguars. The Jaguar Corridor Initiative conservation program seeks to link the breeding populations in isolated areas from Argentina through Mexico and supports mitigating efforts between humans and jaguars.

Believing that jaguars regularly cross the borderland, the US Fish and Wildlife Service established an official habitat nearly 1,200 square miles along the Mexican border in southern Arizona and New Mexico in 2014. Any large or extensive fence or barrier between the countries would prevent a critical exchange of genetics.

LEGALITIES

Jaguars are protected under the Endangered Species Act and listed as Threatened with Extinction by the International Union for Conservation of Nature. All international trade in jaguars or their parts is prohibited through the Convention on International Trade in Endangered Species (CITES).

Dealing with Jaguars

Jaguars are exceedingly rare in the United States. It is highly unlikely for a person to encounter the elusive hunter outside of the very small areas where they have been observed, primarily by cameras at night. Most experts do not believe a breeding population is present in the United States at this time.

Historic range, early 20th century

Current range

▲ The fossil record indicates that jaguars once roamed much of North America. Jaguar fossils commonly turn up in the Southeast; however, skeletons and carved jaguar bones have been unearthed as far north central as Ohio, northeast as Pennsylvania, and northwest as Oregon and Washington.

DAMAGE ID: Jaguar

PREY ON

Small and large livestock

METHOD OF KILL

▸ Prey pounced on from cover or trees and killed with a bite to the throat or skull

▸ Carcass dragged to a secluded location and covered between visits

▸ Neck and chest eaten first, heart and lungs consumed, followed by the shoulder area

▸ Territory marked with circular scrapes of dirt around scat, brush piles soaked with urine, and clawed trees

TRACK

Front 2⅞–4⅞ inches long, 3–4⅞ inches wide; rear 2¾–4 inches long, 2½–4½ inches wide; no claw marks; larger than mountain lion, with broader, rounded tips on the toe imprints

Front

Rear

TIME OF DAY

Nocturnal; most active dawn and dusk

GAIT

Stride at walk 16–32 inches or more, direct-register or overstep

SCAT

Blunt or tapered ends, 5 inches long and 1½ inches in diameter, often wrapped in fur

5"

BOBCAT
(*Lynx rufus*)

Although most people have not seen a bobcat, at least 1 million of these North American wildcats live in the United States. Indications show that the population is increasing in most areas, and their range is expanding. Bobcats are highly adaptable, living in deep forests, wooded or brushy farmlands, mountains, scrublands, deserts, and swamps. They are increasingly seen in suburban and urban environments as well, being opportunistic and flexible in diet, unlike their cousins the Canada lynx. The bobcat's modern success in adaptation is similar to that of the coyote. When it is sighted, however, the bobcat is often confused with a feral cat, a mountain lion, or a lynx.

Widely dispersed across much of the continent and successfully expanding its range, the bobcat is secretive and rarely seen.

Evolved from the Eurasian lynx, which migrated to the Western Hemisphere about 1.8 million years ago, the species was present as the modern bobcat by 20,000 years ago. Originally, the animal was called the *wild cat* by the European colonists; the common name *bobcat*, for its bobtail, did not appear until the 19th century. The earliest settlers hunted and trapped bobcats for fur and sport. Even today some 30,000 pelts are exported from the United States each year, down from a high of 90,000 pelts in the late 1970s.

DESCRIPTION

A bobcat is usually taller and twice the size of a domestic cat, with longer hind legs, a distinctive ruff, and a bobtail. It is usually smaller than a mountain lion, is differently colored and patterned, and lacks the long tail. Compared to the rarer lynx, the bobcat is usually shorter and stockier, with much smaller feet. It is more heavily boned and muscular than a large domestic cat, even when similar in size.

The bobcat has excellent sight, hearing, and smell. It uses its retractable claws to climb trees to escape threats and can swim if necessary. The hind legs are longer than the front legs, but not so long as on the lynx.

Size. Bobcat size varies by location as well as by sex, with males significantly larger than females — from 25 to 80 percent larger. Bobcats can range from 26 to 41 inches long, 20 to 24 inches tall, and 11 to 40 pounds; however, on average a bobcat will weigh around 20 pounds. Their distinctive bobtail is 4 to 8 inches long.

Coat and coloration. Varying according to habitat, bobcat coloration provides excellent camouflage. A bobcat may be light gray, yellowish, or reddish brown, with dark brown to black stripes and spots, a dark line running down the back, and bars or stripes on the front legs and the tail. The back of the ears is black, as is the top of the tail tip. The chin, throat, and underparts are lighter. The ears have slight tufting. The facial ruff makes the face look round, although it is noticeably smaller than the distinctive ruff of the lynx.

As many as 13 separate subspecies exist, although taxonomic disagreements exist over the exact divisions. Bobcats certainly have varying size, coat color, and spotting patterns. They tend to be lighter in deserts and darker in northern forests. Bobcats are more yellow or red in the southern states, with a few rare black bobcats in Florida. Longer hair is found in cold climates. Bobcats are usually larger in the West and in the North, and smaller in the South.

Lynx–bobcat hybrids are sometimes located where their habitats overlap, such as Minnesota, Maine, and New Brunswick in Canada. These offspring are fertile and possess larger ear tufts than the bobcat, but smaller feet and bobcat-like coats.

HABITAT AND BEHAVIOR

Bobcats are found in many diverse areas, including forests, grasslands, scrublands or deserts, canyons, wooded streams or river land, mountains, swamps, subtropical forests, and farmlands with nearby dense cover. They are increasingly seen in suburban and urban areas where they can obtain water, food, and shelter. They are generally absent from large urban areas on the coasts and from areas in the Midwest where agricultural fields are large and lack sufficient shelter or cover. They are not adapted to deep snow like the broad-footed lynx, which limits their range. With climate change and land clearing, the bobcat is moving into some areas formerly held by the lynx.

20-24 inches at shoulder

Tail 4-8 inches   26-41 inches from tip of nose to rump

▲ Bobcat

Bobcats use rock ledges, caves, brush piles, hollow logs, and abandoned animal dens for shelter, often maintaining several spots in their range. They are territorial, marking their individual ranges with scent and claw marks. Males can be somewhat tolerant of nearby males. While several female ranges may overlap a male's range, females do not associate with one another. Bobcats will use animal and human trails as well as remote roads for travel within their home range. Range sizes vary with climate, habitat, availability of food, and predator threats.

Hunting and foraging. Bobcats are solitary but less reclusive and more aggressive than lynx. They begin to hunt well before dusk and can continue into the morning after sunrise, but they can also be out during the day, especially during fall and winter, in search of food. They hunt primarily on the ground, both ambushing and stalking their prey. Although fast sprinters, they are not distance runners, pouncing on their prey and killing with a bite to the neck.

They prefer rabbits and snowshoe hares but also hunt squirrels, rodents, small mammals, and birds including large turkeys, geese, swans, or cranes, and also nestlings or eggs. They occasionally hunt deer, usually waiting until the deer are bedded down. Bobcats will infrequently kill small livestock, rabbits, piglets, poultry or other birds, and domestic cats or small dogs. They will also scavenge carcasses and opportunistically eat small predators, fish, reptiles, amphibians, and insects. They cover larger kills or scavenged carcasses with leaves, debris, or snow and return to feed again several times. They will aggressively defend a carcass.

LIFE CYCLE

Breeding season can vary in different locations but usually occurs in late winter or early spring. Males and females can have multiple partners, and when breeding they will yowl and scream like domestic cats. About 60 days later, 1 to 6 faintly spotted and streaked kittens are born. Kittens usually stay with their mother for at least 8 months. At times, litters can arrive as late as fall. Females can breed the following year, with males generally a year later.

In the wild, bobcats can live 12 years or more. Other predators, including other bobcats and raptors, prey on young bobcats. Mountain lions, wolves, coyotes, or other large predators will kill adults. Road accidents, starvation, disease, and poisoned baits also take their toll, but half of adult bobcat deaths can be attributed to hunting or trapping.

HUMAN INTERACTION

Bobcats are an important predator in the ecosystem and are especially useful in the control of rabbits,

▲ Bobcats are found throughout most of the United States and across southern Canada. The density of bobcats is higher in the southeastern states. They are absent or rare much of the Midwest, nearby areas, and portions of the Mid-Atlantic states. Bobcats are moving northward with warmer winters and habitat changes, including into northern Minnesota and southern Ontario.

rodents, and pest species such as cotton rats. When bobcats prey on livestock, sheep, goats, and poultry are the most common victims, but they will also take calves, domestic rabbits, piglets, other birds, and eggs. Bobcats will also prey on domestic cats, small dogs, and exotic or pet birds.

LEGALITIES

Bobcats are protected as Endangered in Indiana, Iowa, and New Jersey. Hunting or trapping is regulated in most states and Canada, although some states have continuously closed hunting seasons. Some states allow landowners to kill a bobcat without a permit. Check with state or provincial authorities before dealing with a problem bobcat. Report potentially dangerous bobcats to local authorities for removal. Keeping a bobcat as a pet is illegal in many states or requires a permit.

Bobcat Encounters

Rarely seen, bobcats behave cautiously around humans. They will growl, hiss, or spit if threatened or protecting a kill. Bobcat attacks on humans usually involve rabid or unusually aggressive individuals.

Open all gates and remove dogs if a bobcat is in a yard. If the bobcat is up a tree, it may have been chased there by a dog or may be just resting. Stay indoors, remove pets from the area, and contact authorities.

Do not pick up or approach a bobcat kitten, despite its adorable appearance. Even if you do not see her, the mother will be extremely protective if she perceives a threat to her offspring. If a bobcat has kittens in a den on your property, exercise caution and keep people and pets away. The mother may be encouraged to move her kittens if you play continual loud noise or flashing lights, although she will be reluctant to move them if they are very young. Seek assistance if necessary.

▲ Avoid surprising a bobcat by making noise when walking on trails, especially around blind corners or in known bobcat habitat. Remain observant for signs of bobcats — claw marks, scat, or tracks. Carry an air horn or bear spray when hiking or walking pets in bobcat territory.

If You Meet a Bobcat

- Do not approach the animal.

- Do not run, but do back away slowly from the bobcat or its kill.

- Pick up and protect small children.

- Make loud noises, use air horns, or spray water, if available.

- In the rare case of an attack, the bobcat will target the head, neck, or shoulders. Use pepper or bear spray, fight off and hit the cat, try to protect your head and neck, and make loud noises.

Dealing with Bobcats
Homes and Yards

In suburban or urban areas, you may need to work with your neighbors to discourage bobcats. Be aware that roads or trails through your property may be used by bobcats to access prey.

- Eliminate all attractants, including water in arid areas.

- Do not feed wild birds or animals in your yard.

- Do not feed bobcats.

- Do not leave pet food or water outside.

- Clear brush and other hiding places in your yard and around buildings, cover access under decks or porches, and secure buildings and animal housing.

- Keep backyard poultry or other pet birds securely penned; the pen must have a secure top, not a tarp or flimsy netting. Cats and dogs may need similar protection if left alone in the yard.

- Use LGDs, which provide excellent protection in larger yards.

Livestock Husbandry

- Keep vulnerable animals in more secure and protected pastures or paddocks, for example, during livestock birthing seasons.

- Remove sick or injured animals and carcasses.

- Use electronic motion-sensor guards with sounds or lights.

- Play loud music or shine continuous bright lights to temporarily discourage bobcats.

- Maintain a human presence to discourage bobcats.

- Use LGDs, which provide excellent protection.

Fencing

Bobcats can easily jump 6 feet, and may jump up to 12 feet.

- Make sure that fences are tight and secure.

- Place electric scare wires outside the fence at 12 and 18 inches above the ground with another near the top of the fence.

- Add an overhang, extending outward from the top of the fence.

- Place metal or plastic guards at least 5 feet high to discourage bobcats from climbing wooden posts and other structures, trees, and overhanging branches to gain access to pens or yards.

- Construct pens from woven wire or multiple stands of electric fencing, including a sturdy roof or cover.

DAMAGE ID: Bobcat

PREY ON

Small or young livestock, poultry, pets

TIME OF DAY

Before dusk through early morning; daylight hours during fall and winter

METHOD OF KILL

A field autopsy is important to determine whether the animal was killed or just scavenged by a bobcat. It can be difficult to distinguish between cougar, coyote, or fox kills when a carcass is discovered.

▸ Poultry dead, head and neck bitten; shoulders bitten on large birds; head eaten

▸ Small animals killed by a single bite to the head, neck, or throat, killing the animal through suffocation or shock; jugular vein often punctured; bite mark with a ¾– to 1-inch space between the canine teeth, smaller than a mountain lion's bite; claw marks on the shoulders, back, or sides that leave hemorrhages

▸ Adult animals usually attacked when bedded down, seized by the neck, then bitten on the back of the skull, throat, or front of chest; claw marks visible on the back, shoulders, or sides

▸ Animals dragged or carried away; small animals and birds eaten completely

▸ Larger animals eaten under the ribs, beginning with viscera but leaving the rumen, or from the neck, shoulders, or hindquarters. Both bobcats and mountain lions leave cleanly cut edges on bone and tissue, compared to the ragged edge left by canines.

▸ May return to larger kills several times, moving the carcass and covering it with soil, debris, or snow, leaving marks showing a 15-inch reach (compared to mountain lions' 24-inch reach)

▸ Once bobcats find prey at a location, they will return to seek more.

TRACK

Front 1⅝–2½ inches long, 1⅜–2⅝ inches wide; rear 1½–2¼ inches long, 1¼–2⅝ inches wide. No visible claws; 4 toes and pad visible.

Front

Rear

SCAT AND SIGN

Blunt, dry, 3 inches long, and ¾ inch wide. Kittens cover their feces, but adults leave feces uncovered.

3"

GAIT

Walking stride about 12–18 inches, overstep or direct-register. Running stride up to 4 feet.

OTHER SIGN

Claw markings 10–12 inches long on trees

CANADA LYNX
(Lynx canadensis)

Easily confused with the bobcat, the other member of the lynx family in North America is the Canada lynx. Genetic evidence suggests the Canada lynx is actually descended from a much more recent migration of the Eurasian lynx into North America. It is more commonly just called the lynx, and it has evolved to be a highly specialist predator of the snowshoe hare (*Lepus americanus*). The hare forms the majority to the entirety of the lynx's diet when available. In Alaska and Canada, the snowshoe hare has an approximately 10-year-long population cycle building up into a boom followed by a bust. The crash in snowshoe numbers dramatically affects the lynx population as well.

Black tufted ears and a long, pointed facial ruff help distinguish the lynx from the bobcat.

The lynx now occupies the portion of the snowshoe's range where it has little or no competition from other predators — the dense forests of interior Alaska and Canada. The historic range of the lynx extended south into 24 US states across the North and into the Rocky Mountains. In the 19th century, habitat loss and trapping reduced the population in many areas, although it remained stable in more remote terrain. Again in the 1970s and '80s, high prices for lynx pelts raised growing concern in terms of species vulnerability. In 2000, the Endangered Species Act listed the lynx as Threatened in the lower 48 states. In 2014, critical habitats were designated in Maine, Minnesota, Montana, Idaho, Washington, and the Greater Yellowstone area in Wyoming. Washington and Montana have the largest populations outside of Alaska and Canada.

DESCRIPTION
Size. Smaller than the Eurasian lynx, the lynx is typically larger than most bobcats. Ranging from 10 to 38 pounds, Canada lynx average 22 pounds, with males larger than females. Standing 19 to 22 inches tall, lynx can be from 24 to 42 inches long.

The lynx has a small head and very long legs. The hind legs are longer than the front, much like a jackrabbit's. The large, broad paws, equipped with retractable claws, are well suited to deep snowpack. The lynx is a good climber and swimmer, able to cross large rivers. It has excellent hearing and vision. Of the 3 subspecies, the Newfoundland lynx is large enough to kill caribou calves.

Coat and coloration. The lynx has a long, thick coat in winter, colored silvery or grayish brown and turning more reddish or yellowish brown in summer. The dark brown or black spots or small stripes are less defined than on the bobcat. White tips on the hairs give the lynx a frosted appearance. The lynx also has a very distinctive double-pointed long facial ruff and long black tufts on large, triangular ears. The ruff is as sensitive as the whiskers. The tail is shorter and blunter than that of the bobcat, colored brown to whitish buff with black hair encircling the tip. The blue lynx, a rare genetic mutation, is a very pale color.

A lynx should appear twice the size of a domestic cat, closer in size to a medium-sized dog. Lynx appear grayer than bobcats, with their very distinctive facial ruff, ear tufts, and black-tipped short tail.

HABITAT AND BEHAVIOR

The lynx lives primarily in cool, mature coniferous forests with dense undergrowth, deep snow, and fallen timber. It may hunt in younger forests with thick vegetation and in rocky areas or tundra but is less common in mixed forests or wooded farmland. Lynx shelter under rock ledges or fallen timber in dense areas of the forest. Females choose a birthing site in thick brush, thickets, woody snags, or hollow logs and do not dig dens. Loss of timber, both mature trees and fallen logs, negatively affects the lynx, as do roads for logging or recreation, which allow other predators increased access to the snowshoe hare.

The range of a male lynx is usually less than 20 square miles but will increase when prey is scarce and the cats roam even farther in search of food. Female territories may overlap each other and that

19–22 inches at shoulder

24–42 inches from tip of nose to rump

▲ Canada Lynx

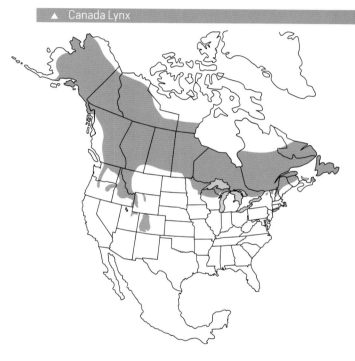

▲ Lynx are most common in the boreal forests of Alaska and Canada. A few hundred lynx are found in the lower 48 states, primarily in Montana, Idaho, Washington, and Wyoming around Yellowstone National Park and Medicine Bow National Forest; more than 1,000 are now found in Maine. Small breeding populations are found in Vermont, New Hampshire, New York, Minnesota, Wisconsin, the Upper Peninsula of Michigan, Oregon, and Utah. Lynx were extinct by the 1970s in the Colorado Rockies; however, 204 lynx were successfully introduced. Lynx have also been observed in Iowa, Nevada, New Mexico, and North and South Dakota. Habitat loss, fragmentation of populations, and reduced snowpack all negatively affect their population.

of the male. Young females usually occupy ranges near their mothers, and adult mothers and daughters may continue to travel or hunt together, although lynx are usually solitary hunters. Generally nocturnal, lynx are most active during dusk and dawn, and they may hunt or travel during the day when necessary. They can hunt in deep snow and at higher altitudes, both walking on the ground and on fallen timber. Lynx either ambush their prey or actively search and flush them from cover. Lynx are very fast at a sprint but do not have stamina and will only chase prey a few dozen yards.

In the absence of their preferred prey, lynx will hunt small- to medium-sized animals such as squirrels, rodents, and ground-nesting birds. In fall and winter they will occasionally take hoofed mammals. They also scavenge carrion. In the southern areas of their range, their diet is likely to be more diverse.

LIFE CYCLE

Lynx breed in winter or early spring. Females mate with only one male, although the male may breed several females. About 64 days later, 1 to 6 kittens are born and become fully independent 10 months later. Young females breed when they are 1 or 2 years old, with males waiting another year. In the wild, lynx can live 14 to 16 years. Secretive and reclusive, the lynx is less confrontational with larger predators compared to the bobcat. While young lynx fall victim to larger predators, both adults and young are lost to starvation, road accidents, or trapping.

Lynx–bobcat hybrids are sometimes identified near the limits of the lynx range in Minnesota, Maine, and New Brunswick in Canada. These offspring are fertile and possess large ear tufts but smaller feet than the lynx and have bobcat-like coats.

HUMAN INTERACTION

The lynx is an important predator of the snowshoe hare. Lynx rarely kill sheep, goats, domestic rabbits, cats, poultry, or game birds. The lynx is trapped for fur in Alaska and most of Canada.

LEGALITIES

Outside of Alaska, lynx are protected as a Threatened species in the United States and as Endangered in Nova Scotia and New Brunswick. Regulated trapping seasons are allowed elsewhere in Canada and Alaska.

Lynx Encounters

In the United States, an encounter with a lynx would be extremely rare outside of their known range. The secretive feline may be spotted near the forest edge but is rarely seen. The only two reported attacks on humans involved people carrying dead snowshoe hares or wearing buckskin; both were believed to be a case of mistaken identity by the lynx.

Dealing with Lynx
Homes and Yards

- Clear brush and fallen trees, since the lynx uses them for hunting.

- Play loud music, shine irregular or bright lights, and produce other sounds as good frightening techniques from dusk through dawn, since lynx avoid all human activities.

- Use LGDs; their barking is a good deterrent. Lynx are not usually aggressive toward dogs.

Livestock Husbandry

Predation on sheep, goats, calves, or poultry is more likely in the fall and winter or when the cyclical snowshoe hare population is low. Keep stock in safer areas during these times. Clear brush and fallen trees away from animal enclosures and paddocks.

Fencing

Woven wire should protect the sides and tops of pens or coops, since lynx can climb wooden fence posts and can jump 6 feet high.

DAMAGE ID: **Canada Lynx**

PREY ON

Sheep or goats, rabbits, poultry, domestic cats

TIME OF DAY

Nocturnal, most active dawn and dusk

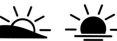

METHOD OF KILL

Lynx will opportunistically feed on carcasses of dead stock. A field autopsy is essential before assigning blame to a lynx found feeding on a carcass.

▸ Poultry killed less often by lynx than by other animals

▸ Birds killed by a bite on the head or neck; head eaten; eggs eaten

▸ Small stock, rabbits, and domestic cats killed by a bite through the top of the head or neck by the lynx's 4 long canine teeth

▸ Larger animals jumped on, their shoulders or back attacked, with claw marks left on their sides, their throat bitten; killed by suffocation

▸ Hindquarters usually eaten first, along with shoulder, neck areas, and flank

▸ Larger prey or a scavenged carcass dragged and cached under brush or rocks, covered with leaves, scratched out 12–14 inches.

▸ May kill several smaller animals and cache them

TRACKS

Front 2⅜–4¼ inches long, 2⅜–5⅝ inches wide; rear 2½–4⅛ inches long, 2⅛–5 inches wide; often indistinct due to heavy fur, small size of pads, or in deep or fluffy snow. In fluffy snow track may leave a uniquely cross-shaped mark. No claw marks.

Front

Rear

GAIT

Walking stride 11–19 inches, overstep or direct-register

SCAT

Blunt, dry, 4½ inches long, and ¾ inch in diameter; not covered

4½"

OTHER CATS

Jaguarundi
(*Puma yaouaroundi*)

Despite the significant difference in size, the jaguarundi is closely related to the mountain lion. Its range extends from Brazil and Argentina, north through Central America and Mexico, and into the United States. In the United States, the extremely rare jaguarundi has a protected area at the Laguna Atascosa National Wildlife Refuge, although its presence has not been confirmed in Texas since 1986, when a dead jaguarundi was found on a roadside. Other sightings have occurred in southern Arizona; the area of Big Bend, Texas; Florida; and coastal Alabama. The small population in Florida may be an introduction. It is protected as an Endangered species in the United States, although unthreatened elsewhere in its range.

Only a bit larger than a house cat, the jaguarundi is somewhat weasel-like in its facial appearance, giving rise to its nickname the *weasel cat*. It has a long body, shorter legs, and a long, somewhat flattened tail. It weighs from 7 to 20 pounds and ranges from 21 to 30 inches long, with a 13- to 23-inch-long tail. Its head is small and flat, with a weasel-like nose; small, rounded ears; and somewhat narrow eyes. Jaguarundis come in 2 uniform colors — shades of gray and red. Light tawny-red cats can resemble a small mountain lion.

The solitary and shy jaguarundi is active in the day, hunting on the ground for small mammals, reptiles, and birds. Adaptable to a variety of habitats from dry to wet, jaguarundis seem to prefer forest edges and brushy lowlands. They are capable of reproduction year-round. The tracks and scat are very similar to those of the domestic cat.

For information on domestic cats, see pages 181 to 184.

Ocelot
(*Leopardus pardalis*)

The beautiful ocelot once roamed throughout much of Texas, east to Arkansas and Louisiana, and west into Arizona, but it is now primarily found in the United States in southern Texas, where it is protected at the Laguna Atascosa, Lower Rio Grande Valley, and Santa Ana National Wildlife Refuges. Unfortunately, with fewer than 100 animals surviving in Laguna Atascosa, vehicles have struck several ocelots in recent years. In 2016, wildlife underpasses were constructed under roads around the refuge and efforts began to establish safe corridors for travel on nearby private land. A single ocelot was recently observed in Arizona. The ocelot is protected as an Endangered species in the United States, with a population estimated at 80 to 120. Ocelots in Texas may belong to 2 separate subspecies. Elsewhere, the ocelot's range extends down the coastal areas of Central America to South America, where they are most numerous in the Amazon basin.

Ocelots are considered medium-sized felines, more slender than a bobcat, weighing 20 to 35 pounds, and ranging from 37 to 50 inches long. The legs are sturdy, the tail is short, and the ears are rounded. The stunning coats are uniquely patterned in each individual. The base color is tawny yellow to red to gray, with lighter underparts. The pattern is composed of brown rosettes with black borders, black spots, and stripes on the head, body, legs, and tail.

The ocelot is considered one of the most abundant and successful smaller cats. Ocelots prefer dense forests or cover, although they are found in varying habitats. Usually nocturnal, the solitary ocelot tends to sleep in trees where it can also capture monkeys, but it most often hunts on the ground for small mammals, rodents, reptiles, birds, and fish. Ocelots are capable of year-round reproduction. Despite an international trade ban on spotted cat pelts, ocelots remain vulnerable to habitat loss, illegal hunting, and roadside accidents. The tracks resemble those of the bobcat, about 2 inches across.

Bears

Ursidae

Bears are "dog-like" carnivores that usually have nonretractable claws and long snouts. Eight species of bears are found in North and South America, Europe, and Asia. Large, powerful, and imposing, bears have been both hunted and revered by humans. Three species of bears are found in the United States and Canada — black, brown or grizzly, and polar bears. Migrating from Eurasia, both brown and black bears evolved separately from a distant ancestor some 5 million years ago.

OVERVIEW. North American bears have dense, long hair; large heads and bodies; long snouts; short tails; strong, stocky legs; and paws that walk **plantigrade**, or flat on the ground. Males can be up to twice as large as females, but both are the strongest large animal in North America, capable of lifting, moving, or dragging objects much heavier than themselves. They also deliver powerful blows with their front legs.

Many bears are excellent climbers, at least while young, and they all swim well. Although they walk slowly, bears have a fast and silent run. Bears can stand on their hind legs to walk a few steps, to reach higher for food, or to fight. Their most important sense is smell, which is 7 times better than that of a scenting breed of dog. Bears also have good hearing but poorer vision. Since they are nearsighted, bears probably do not see objects in detail at a distance. They often stand up to look at things in the distance, trying to catch a scent to aid in identification. Bears are widely recognized as intelligent, clever, and adaptable.

Varying in their diet according to habitat, bears are opportunistic feeders and roam constantly through their range in search of a wide variety of seasonal foods. About 85 percent of a black or a brown bear's diet is plant foods, including grasses, sedges, shoots, buds, mosses, fungi, flowering plants, fruits, and nuts. They dig for roots, tubers, insects, grubs, worms, larva, and burrowing animals. Bears also strip trees of bark or tear apart logs for sap, insects, bees, and honey. They fish for salmon, trout, or bass and forage for eggs, nestlings, or birds. Bears scavenge carrion, and winterkills, or steal carcasses they can smell from 2 miles away. When they do hunt, their prey is generally small animals such as rodents or ground squirrels; newborns, fawns, or elk calves; new mothers; the infirm; or floundering animals. More rarely, black or brown bears prey on adult animals or livestock.

Bears are usually active during the day, but proximity to humans will lead them to forage more during the night or around dusk. This pattern also varies by season, especially in late summer and early fall, when bears need to forage up to 20 hours a day to increase their body weight in preparation for hibernation. Driven by hunger, both before and after hibernation, bears are less predictable toward humans they encounter during these times. Bears rest in dense vegetation, in a tree, or by a rock. On the ground they leave a slight roundish depression, often with some vegetation or branches on the ground in cold weather.

Generally solitary except during breeding season or while raising cubs, some young adults will band together for hunting or playing. Adult bears will attack smaller bears and each other at times. Bears do gather near exceptional food sources such as landfills and spawning salmon. The largest and most dominant bear will defend the best areas, and a hierarchy will develop.

LIFE CYCLE. During breeding season, both sexes will mate with more than one partner, and cubs in the same litter may have different fathers. Males will attempt to defend a female from other males but will then leave her to bear her cubs alone. In female bears the embryo remains dormant, a process known as **delayed implantation**, until fall when she hibernates. About 3 months later, while she sleeps, 2 to 3 cubs are born weighing only about 1 pound. This young-to-adult size ratio is the smallest of any placental mammal. Helpless and nearly hairless, the cubs wake, nurse, and grow until they emerge from the den with their mother later in the spring when the snow melts. The cubs remain with her until the next spring or even after their mother's breeding the next summer until she forces them out of her territory. Generally female bears have litters 3 to 6 years apart, a very low reproductive rate.

The form of hibernation in bears is more accurately called **winter dormancy**, adapted to meet the challenges of cold, deep snow and low food sources during winter, as well as immature cubs; but bears remain alert enough to be roused if needed. Bears do not eat, drink, or eliminate during this time. Not all bears hibernate, however, and in those that do, the length of hibernation depends on climate, food sources, and sex.

The illegal sale, primarily from poaching, of North American black and grizzly bear parts to Asian markets has increased even as the supply of Asian bears has decreased. Desirable parts include blood, bones, claws, paws, fat, meat, hides, teeth, and gallbladders.

AMERICAN BLACK BEAR
(*Ursus americanus*)

On the East Coast of North America, early explorers and colonists noted the abundant presence of bears. Black bears were found almost continent-wide, except the deserts of the Southwest and far northern Canada. American black bears are related to Asian black bears, but they are a separate species found only in North America. Regionally, the 17 black bear subspecies are often known by local names, such as the Florida, California, and Olympic black bears, or by colors, such as the cinnamon bear or the rare, blue-gray glacier bear. Worldwide, black bears are more numerous than all other bear species combined.

Naturally nearsighted, bears stand on their hind legs to catch the scent of something in their distant vision.

Since European explorers and colonists were well acquainted with bears, when the black bear was encountered he was simply called *bear*, an Old English word, and sometimes *bruin*, which was French. As the native people did previously, colonists widely hunted bears. Bear meat was commonly eaten, and the fat, or bear *grease*, was used for cooking, medicines, cosmetics, and waterproofing. More valuable than brown bearskins, black bear fur was desirable for its qualities of softness with a dense undercoat and longer, coarser outer hair. Despite the reality of the bear as a powerful predator and the horrors of bearbaiting, bears would come to be favorite animals in circuses, films, stories, and sports (as mascots).

Despite their powerful presence, bears are favorite fictional characters — including Winnie the Pooh, the original teddy bear, and Smokey Bear.

At the time of European colonization, the black bear population was estimated at about 500,000. Hunted extensively and faced with continual loss of their favored forested habitat, black bear populations were seriously reduced in most areas and exterminated in others. And yet today the numbers of black bears have actually increased to as much as 900,000. They are still absent from much of the center and southern areas of the country because of the lack of their preferred forest cover. However, they are again found in most states, although sometimes just as transients or in small isolated populations in national parks or recreation lands. Today their range overlaps with brown or grizzly bears in areas of the western United States and Canada. More bear–human interactions occur as people encroach on bear habitat for both recreation and housing. Unfortunately, bears easily adapt to human-provided food.

DESCRIPTION

Smaller than either brown or polar bears, the average adult male black bear is 4 to 6 feet long, standing 2.5 to 3.5 feet at the shoulder, and weighing 250 to 400 pounds, with some individuals as high as 600 pounds. Females are smaller, averaging 150 pounds. The bear's moderate-sized head has small eyes and round, erect ears, with a slightly roman-nosed and tapering muzzle. In profile, the black bear lacks the prominent shoulder hump of the grizzly and has a 3- to 6-inch tail. Black bears also have shorter, curved claws about 1 inch long.

East of the Mississippi River, 70 percent of the population is colored black. The muzzle is lighter brown or buff, and a white blaze on the chest is usually present at birth, although most fade away except in the Florida subspecies. About half of the populations of black bear subspecies in western areas are light to dark brown, cinnamon, or blond. In some specific areas, 95 percent of black bears are actually brown. Cubs can be different colors in the same litter, and their colors may change as they age, between shades of black and brown. Although rare, albino black bears also occur.

Black bears are excellent climbers and will climb trees to eat, to reach safety, and sometimes to hibernate. They climb less often when older and heavier. Black bears have a true walk, not a shuffle, and can run 25 to 30 mph for short distances. They are dexterous and agile, often seen walking on narrow logs. Black bears can pick up very small objects, as well as open screw-top jars and door latches.

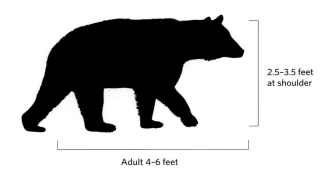

2.5–3.5 feet at shoulder

Adult 4-6 feet

▲ Black Bear

HABITAT AND BEHAVIOR

Black bears prefer dense forests with understory brush, and they are found in hardwood or mixed nut trees, wet or dry forests, hardwood swamps on the eastern coast, coniferous forests of the Rockies and western coast, and chaparral or pinyon in the Southwest. In mountainous terrain, bears will move up and down slopes foraging in different seasons. They prefer remote or wilderness areas but will move out to meadows, timbered land, or other open areas in search of food. Black bears are also seen on roadsides, at landfills, and in picnic or camping areas.

The range size depends on habitat and food availability. Males can have ranges up to 1,000 square miles, which may overlap against several females with ranges of 6 to 19 square miles. Black bears use trees, poles, or wooden structures, marking with body rubs, shedding hair, claws, and teeth. The reason for these marks on "bear trees" is not completely established, but they may be dominance marks warning other bears away.

Primarily vegetarian and opportunistic, black bears are not active or skilled predators, consuming more carrion than hunted animals. They are more carnivorous in some areas such as the subarctic. They will steal carcasses from smaller predators but tend not to challenge larger brown bears, wolf packs, mountain lions, or wolverines. They may share carrion with a single coyote or wolf. When they do hunt, black bears will ambush or approach on four legs, using a blow from a front leg or a bite to disable or kill their prey. They will drag prey to secluded areas to consume it but may not cover it between feedings.

Black bears hibernate for 3 to 8 months, depending on climate and (for females) pregnancy. In northern areas, hibernation begins October to November and ends April to May. Bears may lose 25 to 40 percent of their weight during this time. They may hibernate in caves, hollow trees, culverts, or buildings, although most dens are dug under rocks or logs and windfalls of timber or brush. In areas without snow, black bears often den higher in tree cavities. In warmer weather, black bears may rouse to forage for food; in southern areas, bears may sleep for only a few days, except for pregnant females, which necessarily sleep for longer stretches.

◄ Approximately half of the North American black bear population lives in Canada except on Prince Edward Island. Approximately 100,000 black bears live in Alaska. In the lower 48 states, black bears are found from the northeast, down the length of the Appalachian Mountains into Georgia, the northern Midwest, the Rocky Mountains, the West Coast, and Alaska. Sizable populations are found in California, Colorado, Idaho, Massachusetts, Michigan, Minnesota, Montana, North Carolina, Oregon, Pennsylvania, Washington, and West Virginia. Isolated populations are found in Atlantic coastal areas, Arkansas, Missouri, New Mexico, and Arizona. Transient bears occur in Kansas, Nebraska, Rhode Island, and South Dakota. Black bears are not found in Delaware, Illinois, Indiana, or Iowa.

LIFE CYCLE

Breeding season is usually June to July, with cubs born in January to February. Black bear cubs weigh 40 to 60 pounds at 6 months. They reach sexual maturity at about 3.5 to 4 years. Females reach full size at age 4 or 5, which is sooner than males, which grow until they are 10 or older. The average life span in the wild is 18 years, but males can live to 30 and females as long as 40 years. About 60 percent of cubs survive their first year, with the losses due to other bears and predators, including birds of prey. As adults, predation by a grizzly bear or by wolf packs can occur. Ninety percent of adult deaths are attributable to hunting, trapping, or poaching; removal of nuisance animals; road accidents; or accidental poisoning.

HUMAN INTERACTION

Black bears are easily conditioned to feed at garbage or trash areas, landfills, campsites, or direct human handouts. Since feeding causes increased aggression, the only response is removal. Relocation often doesn't work, although "hard" releases with Karelian bear dogs and aggressive tactics seem to work better. Most removals actually result in the bear's death.

Black bears also cause extensive damage to cabins, crops, beehives, timber, and orchard trees. They may challenge fishermen or hunters for food and steal carcasses from hunters. They will also eat birdseed, crops, and orchard and other fruits.

Black bears occasionally kill livestock: newborns or young animals, adult sheep, goats, hogs, and occasionally cattle or horses (which may only be clawed on the sides in escape). Black bears may confront dogs that harass or attack them.

Annual hunting kills at least 30,000 black bears, although bearskins are no longer commercially valuable due to low demand.

LEGALITIES

Regulated and licensed game hunting is allowed for black bears in 28 states and most Canadian provinces. In protected states, authorities will remove problem bears and may provide compensation for damages. Most western states allow predation bears to be killed but may require a permit.

Black Bear Subspecies

Glacier bear (*U. americanus emmonsii*)
The glacier bear differs from other black bears by its blue or silvery gray color on the shoulder, back, and flank. They are primarily found in the coastal areas of southeastern Alaska, including Glacier Bay National Park. No population estimates are available.

Kermode bear (*U. americanus kermodei*)
The Kermode, or "spirit bear," is found in the central and northern coastal areas of British Columbia. The white or cream color is found in about 10 percent of the population, while the rest are black. About 500 Kermode bears exist in the area, and they are fully protected.

DAMAGE ID: American Black Bear

PREY ON

Livestock, poultry, beehives

TIME OF DAY

Day, night, and dusk near humans

METHOD OF KILL

- Hair, tracks, scat usually found in area of damage; vegetation smashed down around kill; distinctive musky odor of bear

- Carcasses showing deep teeth marks ½ inch in diameter on skull or neck and shoulders behind the ears; neck or spine broken; at times, nose or face deeply bitten

- Large claw marks of ½ inch on neck, shoulders, and sides of larger animals, either from striking or straddling

- Body opened, internal organs or udders on lactating females removed and eaten first, then hindquarters and other flesh fed on

- Skin of large prey torn off and inside out; skeleton often left intact and attached to skin; remains not scattered, unlike with wolves and coyotes.

- In open area, kill sometimes dragged into cover and covered with debris, bear returning at dusk to feed even as carcass decomposes

- Where stock is confined or unable to escape, 2 or 3 sheep killed

- Eggs or nestlings stolen

- Beehives and frames broken and scattered, bear returning until all eaten

- Corn or oat crops with large areas of smashed stalks and entire cob of corn eaten

- In orchards, bark clawed off to mark tree and to eat inner bark; branches torn off; bushes smashed

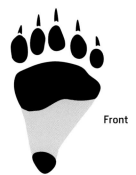

Front

Rear

TRACK

Front 3¾–8 inches long, 3¼–6 inches wide; rear 5⅜–8⅞ inches long, 3½–6 inches wide. Rear foot resembles human footprint, slightly pigeon-toed. All 5 toes, claw marks, and front heel pad may or may not be visible; hair may obscure track details. Can be distinguished from grizzly bear by smaller size, shorter claw marks, and greater curve in arc of toes.

GAIT

Walking stride 17–28 inches, may be direct-register or overstep; lope or gallop can be longer at 24–60 inches.

SCAT

Cylindrical, thick, blunt ends, 7 inches long and 1 inch in diameter, usually coiled, black to brown color. Can resemble human feces except contents may reveal hair, bone, and plant fibers. Looser and softer if feeding on berries. Left uncovered at site of kill.

7"

NORTH AMERICAN GRIZZLY OR BROWN BEAR
(*Ursus arctos horribilis*)

The large family of brown bears (*Ursus arctos*) is found in northern Europe and Asia, although now present only in small numbers. About 50,000 years ago the inland brown bear was believed to have migrated from Siberia into Alaska, later finding its way south into the deserts of the Southwest and Mexico, and east into the prairies of the Great Plains. Historical subspecies included the mainland brown bear, known as the grizzly; the Kodiak and peninsular bears found only in Alaska; and the extinct California (1922) and Mexican grizzlies (1960s). The exact genetic relationships of these subspecies are still being clarified through research.

Grizzly bears can be distinguished from black bears by their large shoulder hump, broad head, concave facial profile, and upturned snout.

The 19th-century population was estimated at 50,000 to 100,000 bears, which was dramatically reduced to some 400 bears in the lower 48 states by the 1970s, occupying less than 2 percent of their historical range. The brown bear population is stable in Alaska, with about 30,000 bears, and in Canada, with some 25,000 bears. In the lower 48 states, a recovering population of 1,500 to 1,800 bears is found in 4 of the 6 protected recovery ranges with possibly very small numbers elsewhere.

In 1804, the Lewis and Clark expedition first encountered the fabled "white bear" in the Dakotas, but the grizzly was well known by the native peoples, who clearly distinguished it from the black bear. The grizzly figured importantly in stories, traditions, and ceremonies; was found on totem poles; and was hunted for food and hides. At times called the Silvertip, grizzlies rapidly gained a fearsome reputation among the European settlers, with much attention paid to the largest examples of the population and sensationalized accounts. Indeed, after many encounters, Meriwether Lewis wrote that the grizzly was a "furious and formidable animal, and will frequently pursue the hunter when wounded." Although one of the most dangerous and unpredictable predators in North America, grizzlies usually avoid human confrontations unless protecting their cubs or food, which they do with great ferociousness. Hunting and settlement reduced the grizzly population, with surviving bears retreating back into the less accessible and less developed areas of their former range. Logging, mining, and further development greatly pressured the survivors beginning in the early 20th century, reducing them to a fraction of their original population and range.

Under Endangered Species Act protection, grizzly recovery areas were established in the North Cascades, Washington; the Selkirk Mountains in Idaho and Washington; Cabinet Yaak in Montana and Idaho; the Northern Continental Divide in Montana; the Bitterroot Mountains in Idaho and Montana; and the Great Yellowstone area across the three states of Wyoming, Idaho, and Montana.

While some populations remain very small, in the Great Yellowstone ecosystem the population has recovered from 136 animals in 1975 to perhaps as many as 700 bears. The Northern Continental Divide population is similar in size and moving out into agricultural and grassland areas. This expansion has prompted a controversial US Fish and Wildlife Service proposal to delist the grizzly bear outside of the national parks and to allow the states of Montana, Idaho, and Wyoming to manage bear hunting. No grizzly bears are present in the Bitterroot range, and the North Cascades population has fallen to fewer than 10 individuals. About 40 bears are found in Cabinet Yaak and another 80 in Selkirk.

DESCRIPTION

Grizzly bears range from 175 to 1,300 pounds, 3 to 5 feet tall at the shoulder, and 5 to 9 feet long. Males average 400 to 790 pounds, and females 290 to 400 pounds. Coastal bears are larger because of a more abundant food source of fish and a longer feeding season. Interior bears in the northern areas are the smallest. The muscular shoulder hump is the highest point in the profile. It can appear even larger with the mane, especially when the hair rises in alarm. The hump is muscular and allows for great strength in the front legs and paws for both digging and hitting. Grizzly bears also have large, broad heads with a dished or concave profile and an upturned snout.

3-5 feet
at shoulder

Adult 5-9 feet

▲ Grizzly Bear

Grizzly bears have a more powerful bite than the black bears. The slender front claws are 2 to 5 inches long and can be used as a tool to remove eggs or roe from a fish and for digging. The tail is 3 to 4.5 inches long.

Silver- or gold-colored tips, or *grizzling*, on the guard hairs on the hump, shoulders, and back gave rise to the name *grizzly bear*. This sheen can create different color patterns on the bears, which are found in shaded colors of pale or creamy blond, reddish blond, light brown, to very dark brown that can appear nearly black. The legs generally appear darker. Darker colors are more frequent in forested areas, with lighter colors in the tundra. Very dark bears are generally not as grizzled. Cubs often have a white collar.

Grizzly bears are extremely strong. They can deliver powerful blows with front legs and paws, drag very heavy weights uphill, move huge rocks and carcasses, dig large holes, and tear down logs or siding. Although generally poor climbers because of their weight and long claws, the cubs or young bears can climb, and the adults can use low branches as ladders to get up a tree. They have a slow, shuffling walk but a fast and silent run of 35 to 40 mph. They also have good endurance and are excellent swimmers.

HABITAT AND BEHAVIOR

Worldwide, brown bears have adapted to a wide variety of habitats. In Alaska and coastal areas, they are found more often near the ocean or rivers. Using dense cover for sheltering during the day, grizzlies also inhabit old and boreal forests, higher elevations, and more open areas such as meadows or even tundra. Males establish ranges from 100 to 1,500 square miles, with female ranges smaller. The largest ranges are in the Rocky Mountains, boreal forests, and arctic areas, with the smallest in coastal areas or on islands. Ranges overlap, and the bears do not actively defend a territory. They do not tend to use the same areas as the black bears.

Grizzlies mark and rub on trees or other objects to communicate their presence, and they also use posturing and vocalizations. They will group together at food sources, defending their dominance, with old males and females with young the most aggressive. Grizzlies will travel long distances for seasonal food sources, which often brings them into increased contact with humans. Grizzlies are also more aggressive than black bears in defense of themselves, their prey, or cubs and are inclined to stand and fight.

◄ The grizzly recovery areas and estimated populations in the United States, shared with Canada in some areas, are in Montana, Wyoming, Idaho, and Washington. Grizzlies are also found in Alaska, often near coastal areas. The Canadian range includes British Columbia, Alberta, the Yukon Territory, Nunavut, and northern Manitoba.

Black Bear or Grizzly?

At a distance bears always appear larger than they actually are because of their dense fur and overall bulky appearance. Excitement can also make identification difficult! Although they may be of similar size or weight, depending on age, black bears and brown or grizzly bears have noticeable differences.

Black

- Smaller and less stocky or bulky
- Rump higher in profile
- Straight face profile and long muzzle, usually colored lighter
- Taller, less round ears, set higher on head
- Short, sharply curved claws 1 to 2 inches long
- Black color east of Mississippi River
- Hind foot track shows spaces between toes and shorter claw marks; front toes more rounded into an arc than the grizzly

Brown/Grizzly

- Larger and more massive overall
- Hump higher in profile
- Very large, broad forehead with dished profile, long upturned muzzle
- Short, rounded ears
- Long, curved claws 2 to 4 inches long, especially on front paws
- Longer and shaggier fur with distinctive mane on back of head and neck
- Blond to dark brown color with frosted hair tips
- Hind foot track shows toes touching each other in a straight line and claw marks extending farther out in front; front track more square than on black bear

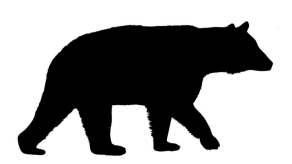

Grizzly bears are omnivores, with their diet often falling between the primarily herbivore black bears and the nearly carnivorous polar bears. Grizzlies will claim carcasses killed by a mountain lion, smaller packs of coyotes or wolves, or a black bear. They generally chase away other predators but occasionally kill wolves, mountain lions, or black bears.

Grizzlies are important predators of elk and moose calves. They will also take large animals such as deer, caribou, bighorn sheep, mountain goats, bison, moose, elk, or muskoxen. Grizzlies often take advantage of animals that are infirm or floundering in snow. They will run down or charge their prey, knocking them over with a blow or blows, and then bite the head or neck. Near the coast, grizzlies eat fish, clams, dead whales, seals, and sea lions. Yellowstone bears have become more predatory with the availability of ungulate calves and adults, bison, winterkilled carcasses, and the occasional sheep or cow. In the Rocky Mountains, grizzlies also eat grasses, legumes, berries, insects, ants and bees, and moth larvae.

Grizzlies will gain up to 400 pounds before a hibernation of 5 to 7 months. They may den as early as September at higher altitudes. Grizzlies often dig a tunnel entrance into a steep sheltered bank or slope to avoid flooding, using the roots of a large rock to help support the sides and roof of a sleeping chamber. Deep snow provides insulation. Grizzlies frequently return to the same area and den. Coastal grizzlies will often hibernate for less time and may use a hollow tree. Grizzlies may not hibernate if food is plentiful during winter, although pregnant females do hibernate. Females will emerge later than males in April or May.

LIFE CYCLE

Grizzlies breed mid-May to mid-July, then go their separate ways. The female dens in late October to November, with the cubs born in January to March.

Weighing 1 pound at birth, cubs will weigh 125 to 175 pounds at 1 year. Eighty percent of cubs survive their first year in Alaska or the Yukon, but in the Yellowstone area subadults have only a 40 percent survival. The young are vulnerable to other bears and natural mortality. Cubs stay with their mother 2 or more years, and she fiercely protects them. Grizzly bear mothers are also known to adopt orphaned cubs. Young females often remain in or near the mother's range, while young males are driven away into less desirable areas. Grizzlies achieve sexual maturity at ages 5 to 6, although males continue growing until age 14. Average life span is 20 to 25 years, but they may live up to age 40. Causes of death for bears over the age of 1 are hunting, poaching, or targeting removal. More rarely, young bears die by road or natural accidents, or attacks by another brown bear, wolf, or mountain lion.

HUMAN INTERACTION

In the lower 48 states, grizzlies are an important keystone predator for ungulate populations. They may visit landfills; damage orchards, beehives, or buildings; and kill livestock. They also challenge hunters or fishermen and damage fishing nets.

Defenders of Wildlife offers funding for electric fencing to protect gardens, orchards, beehives, livestock, and poultry, and to allow backcountry recreation. The group also supports projects to implement bear-resistant garbage and food storage systems in communities and campgrounds.

LEGAL ISSUES

Grizzlies are protected as a Threatened species in the lower 48 states. To shoot them is illegal except in self-defense, in defense of others, or by authorities. Regulated game hunting is allowed in some areas of Alaska and Canada.

Montana will compensate ranchers for verified livestock losses, as the Defenders of Wildlife do in Idaho, Wyoming, and Washington.

OTHER BEARS
Kodiak Brown Bear
(*Ursus middendorffi*)

The largest of the brown bears, the Kodiak may have evolved from a separate migration of Eurasian brown bears into Alaska, where they were isolated since the last Ice Age. They are found today only on the Kodiak Archipelago islands. As of 2005, the population was estimated at about 3,500. Colored blond to dark brown, males can be 8 feet long, stand up nearly to 10 feet, and weigh up to 1,500 pounds, with females 20 to 30 percent smaller.

Polar Bear
(*Ursus maritimus*)

Not only the largest members of the bear family, polar bears are the largest land carnivores. Polar bears are native to the Arctic Circle, living and hunting on the edges of the pack ice through the long winter, traveling up to several thousand miles yearly. Nearly completely carnivorous and strong swimmers, polar bears hunt seals, walruses, Beluga whales, narwhals, small mammals, and birds such as sea ducks. During the months of summer ice melt, many bears are confined to the coastal areas or find resting places farther inland. Although they often do not eat, drink, or eliminate during the summer, males and nonbreeding females do not hibernate. Pregnant females eat through the summer and fall, entering dens in late fall where they remain until late winter or early spring. The dens may be found inland, near the coast, or even on ice floes. Cubs remain with their mothers for about 30 months.

Males average between 500 to 900 pounds, with females about half their size. Large male polar bears can weigh 1,300 to 1,700 pounds. They are massive in appearance, with a long neck and large feet. The dense double coat is white but may appear somewhat yellow, gray, or brown at various times.

Most interactions with humans occur during the summer months. Polar bears will also forage around landfills, construction, or hunting sites. Because polar bears are a protected species, hunting is not legal in Alaska but is regulated in Canada. The polar bear population is estimated at 20,000 to 25,000 worldwide; however, with the continued loss of sea ice due to climate change, experts believe the survival of polar bears is gravely threatened. It is unlikely that the population could find sufficient food by switching to geese or other animals.

Hybrid Bears

In 2006 in the Northwest Territory, a hunter shot a bear that proved to be a hybrid between a polar bear and a grizzly bear, and other hybrids have been confirmed as well. On the ABC Islands (Admiralty, Baranof, and Chichagof), a population of bears has a genetic structure that reflects both brown and polar bear contributions. With the loss of polar bear habitat, more hybrid bears may occur.

While brown and black bears have crossbred in captivity, no evidence supports that this occurs in the wild.

Dealing with Bears
Homes and Yards

Talk to your neighbors and try to cooperatively use prevention techniques. Many municipalities have bear regulations as well. There must be absolutely no food handouts or access to garbage.

If you spot a bear in or around the yard, stay inside, bring pets indoors, and do not harass. The bear will move on unless it finds food. Alert neighbors and local authorities if needed. Report any habituated or problem bears, which may be hazed or removed by authorities. Around homes or cabins in bear country, remain aware of your surroundings and move cautiously at vulnerable times and in areas near cover.

All the prevention techniques used with the more common black bear are applicable for grizzlies.

- Clear 150 feet around human and animal areas. Trim tree branches 15 feet from buildings.

- Locate crop areas 150 feet away from protective cover.

- Harvest ripe fruit and vegetables, clear any fallen fruit from ground, and remove garden refuse.

- Don't plant bear-friendly food plants. Water is also an attractant in arid climates.

- Feed pets inside or bring empty bowls inside after feeding.

- Backyard chickens or pets require secure animal housing, including a top or roof and possibly an electric barrier. Backyard chickens should be closed in at night.

- Don't use birdseed and suet feeders from March through November. Some people bring feeders in at night or make sure they are hung at least 10 feet tall and away from trees or buildings. Equip bird feeders with large spill pans to prevent seed from falling to the ground. Clean any spilled seed.

- Secure pet food and birdseed — don't store on decks or porches.

- Use bear-proof trash containers or keep trash inside a secure building or specially built out-buildings. Use caution with highly attractive-smelling garbage: rinse all used food containers, wrap securely, or freeze and place out right before collection. Be very careful of fish parts and meat scraps.

- Clean and store BBQ grills, outdoor kitchens, and fish-cleaning stations. Outdoor refrigerators and coolers attract bears.

- Spray cans and other food preparation areas with ammonia, bleach, or powerful disinfectant.

- Securely close all windows, doors, garages, and pet doors especially at night. Buy reinforced doors if needed. Use round doorknobs or very secure latches. Bears can easily climb stairs, deck railings, and so forth.

- To prevent climbing, use sections of 1 × 4 boards, 4 feet long, equipped with protruding screws or nails every 6 inches. Wire at least 4 of these boards around any tree to prevent climbing.

- Do not store food or trash in vehicles, which bears can open or damage.

- Use night lights, strobes, motion-detection lights, or loud sounds. Move scare devices every other day to prevent habituation.

- Use livestock guardian dogs (LGDs). Bears do not habituate to loud barking dogs.

- Take special precautions with cabins: reinforce windows and doors; clean out all food and scented items or store in bear boxes; disinfect outside cooking areas and tables. Prevent ants or bees from nesting in siding to avoid property damage.

- Avoid scented items. Bears are attracted to certain aromas: cedar, pine, tarpaper, petroleum products, starter fluid, turpentine, kerosene, and citronella.

Livestock Husbandry

Most livestock predation occurs in spring after bears emerge from hibernation and stock is giving birth.

- Securely confine or protect animals giving birth.

- Remove or bury all afterbirth and carcass materials.

- Locate poultry and livestock housing, beehives, and pet kennels 150 feet from woods or areas of cover. Build securely and surround with an electric barrier.

- Place unwelcome mats either studded with nails or electric in front of windows and doors.

- Be aware that an LGD may be able to harass or frighten away a bear, but a bear can easily turn on the dog. Two LGDs are recommended for bear protection.

Fencing

- Electric wires increase the effectiveness of any fence. Teach bears to respect electric fencing by luring them to it with bacon or empty cans that contained fish.

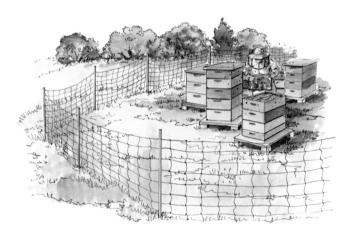

▲ Electric mesh fencing also provides protection for beehives or fruit trees from bears, especially for temporary uses.

- To protect a beehive, orchard, compost pile, or garden, erect an electric fence 3 feet tall, with wires spaced 10 inches apart and a bottom wire 6 inches from the ground.

- Use electric wire or tape fencing to protect poultry or stock.

- Nonelectric fencing should be 6 feet tall — chain link, heavy woven wire, or livestock panels — with a 24-inch outward extension of barbed or electric wire at top.

- If needed, install a 2-foot outward apron of chain link or steel mesh attached to the fence and staked down to prevent digging.

- Group beehives near each other, surrounded by 6-foot-tall livestock panels or 6-inch square mesh fencing, with an additional layer of temporary electric fencing around that. Beehives can also be placed on 15- to 20-foot-tall platforms with support poles wrapped in sheet metal.

ADDITIONAL FENCE CONSIDERATIONS FOR GRIZZLY BEARS

- Temporary electric fencing is recommended for backcountry camping and hunting or stock camps in grizzly country.

- Recommended exclusion fencing for grizzly bears is heavy chain link or woven wire 8 feet high, buried 2 feet belowground, with metal extensions outward at the top with barbed or electrified wire.

- Electric wire fences should be 9 wires, beginning 4 inches from the ground.

- The Defenders of Wildlife will compensate 50 percent of the cost of electric fencing used to prevent grizzly bear access to attractants, including garbage, orchards, or livestock in eligible counties in Washington, Idaho, Montana, and Wyoming (defenders.org).

Portable electric fencing is highly recommended for camping in grizzly bear areas.

Dealing with Bear Encounters

Keep in mind that bears are not generally aggressive toward humans. A bear sighting is also unusual, as most bears are reclusive, although they may be curious. An actual bear attack is extremely rare. Throughout the 20th century, there was an average of 1 fatal bear attack per year, equally due to black and grizzly bears. An increase in fatalities was seen during the 1960s and '70s, mainly due to habituated bears. In the 21st century, fatalities are again increasing slightly, with about 2 deaths per year. Half of these deaths (15) were in Canada, 3 in Alaska, 2 in Tennessee, and single deaths in California, Colorado, Montana, New Mexico, New York, Pennsylvania, and Utah.

If your home or ranch is in bear country, you should be prepared for encounters. Otherwise, outside of Alaska or Canada, the majority of sightings are near national parks, recreation areas, campgrounds, or rural homes where bears have become habituated to humans or their food sources. Bears are also more active during droughts or poor forage years and before and after hibernating, as are young males without a home range. True predatory attacks are very rare.

While similarities are seen between black and brown or grizzly bear encounters, some important differences can also be seen. Knowing how to react if you sight a bear prevents the vast number of encounters from becoming dangerous. Several proven techniques and strategies will also help keep you safe.

Black bears tend to be shy or timid; however, you are far more likely to encounter one due to their widespread presence. Generally, interactions occur because a black bear becomes habituated to humans or human-provided food sources. Other encounters occur when bears are strongly motivated by hunger. Female black bears do not protect cubs as strongly as grizzlies but will react to a perceived threat by humans. Fortunately, even in the case of a physical encounter resulting in injuries, 90 percent are only minor bites, scratches, or bruises.

Grizzly encounters are less common in the lower 48 states because of the bears' very low population. Grizzly bears are more unpredictable and aggressive in their response to perceived challenges to their food or protection of their cubs. Grizzlies will also challenge a fisherman or a hunter for their catch or cache. Their larger size is also a factor.

Bear Spray

The single most important defensive strategy in bear country is to carry bear spray.

Use real bear spray, not pepper spray or Mace. Although the use of spray should be your last resort, it is essential that you know how to use it properly. Used correctly, bear spray is effective in more than 98 percent of attacks, although somewhat less effective with black bears than with grizzlies. Shooting with a gun has proved to be less effective than bear spray because a wounded bear is significantly more dangerous. Defense with a gun in a bear attack is effective less than 50 percent of the time.

Cautions

· Try to stay out of the spray when using.

· Handle carefully, as bear spray may cause blindness at very close range.

· Exercise great care around children.

· Keep in mind that you cannot carry large cans of bear spray on aircraft.

Each person in a group should carry a large can or two (with protective zip tie removed but safety clip on) in a holster or belt outside of clothing or pack. Use with 2 hands, when the bear is from 30 to 60 feet away, spraying short blasts in a side-to-side motion. If the bear is very close, aim for the eyes. Aim lower than the bear's head as the spray will rise. If the bear continues to approach, spray again. Know the capacity of your spray and how many 2- or 3-second bursts you have available.

Canisters with inert ingredients are available for practicing. Learn more about using bear spray at the Interagency Grizzly Bear Committee website (igbconline.org).

Bear-Watching Guidelines

Wild bears are one of the attractions of national parks. However, habituation can be dangerous to bears; they will be destroyed if they become aggressive. To safely view bears and not endanger them or yourself, follow these guidelines:

• Park safely.

• Never feed or entice a bear.

• View bears safely through binoculars or telephoto lenses.

• Stay in your vehicle in national parks, unless the bear is more than 100 yards away.

• Remain very close to your vehicle for safety, and do not let children roam away from you.

• If other people crowd, corner, or behave inappropriately to a bear, it may feel threatened and react. Leave potentially dangerous situations. You are responsible for your own safety in bear country.

Recreation in Bear Country

Remain alert to bear signs:

• Tracks

• Scat

• Claw-marked trees, stripped bark, torn-apart logs

• Digging or large overturned rocks, possibly with claw marks

• Areas of trampled ground or brush

• Oval-shaped day beds

• Signs a carcass may be nearby — odor, fish, fur, or meat scraps on the ground; debris-covered humps on ground; or circling birds

▲ Remain alert when you are in bear country. Do not hike alone. Carry bear spray, and make noise.

HIKING

- Do not hike alone. Travel in groups of at least 2 or 3; groups of 6 or more are safer. By far, most attacks are to solitary people.

- Keep children between adults. Teach them that running ahead, lagging behind, screaming, and running are very dangerous.

- Before hiking, consult authorities for the latest bear activity and areas closed for bear management.

- Stay alert for bear signs. Go back or widely detour around any bear signs.

- Make noise, talk, or sing. Carry a safety whistle, air horn, and other noisemakers. Some experts believe bears associate bear bells with food.

- Carry bear spray and a fixed-blade knife.

- Travel in daylight; midday is best. Bears do use trails at night and may during the day as well.

- Travel in the open whenever possible and use care near blind corners, waterways, animal trails, loud rushing water, or dense brush. Avoid berry patches and other food sources.

- Only carry dry, sealed foods and use non-scented products.

- Do not bring dogs. Off-leash dogs entice bears back to humans. Leashed dogs are also a threat since barking dogs are seen as aggressive to bears.

- Do not set down your food, pack, or gear and walk away.

BIKING, JOGGING, SKIING

Moving rapidly and quietly, bicyclists, skiers, and joggers often surprise bears, who may then chase them. Riders on horseback have never been attacked, although bears may spook horses and cause riders to fall. Follow all the hiking guidelines. Avoid dawn and dusk, when bears are likely to be along roads, trails, or streams. Carry bear spray, make noise, and occasionally use safety whistles or air horns. Stay alert, especially on windy days, around blind curves, or in dense vegetation. Don't wear earphones.

CAMPING, FISHING, OR HUNTING

Always check with authorities as to bear presence. Bears may be habituated to campsites, trails, or fishing sites. Choose an open area away from food sources, 150 feet from cover, streams, or possible animal trails or pathways. Avoid any seasonal feeding or denning areas.

- Don't camp where bear signs are present or at dirty sites. Always disinfect tables, benches, and tents.

- Don't sleep outside your tent.

- Use flashlights after dark.

- Carry bear spray at all times or keep very close, even at night.

- Freeze-dried and other nonodorous foods are preferable. Food needs to be well packaged in airtight containers. Wrap or dispose of all food waste appropriately. Do not bury.

- The sleeping area, tent, sleeping bags, clothes, packs should be free from food and odors. Use scent-free toiletries. Handle menstrual products with care and dispose of appropriately. (Some experts recommend the use of tampons only.)

- Cook, clean dishes, store food (human, dog, or horse) and fish or game meat at least 100 feet, preferably 300 feet, downwind from sleeping area.

- Use bear-proof containers, including provided bear boxes or storage, or hang food. Food should be suspended 10 feet high and 4 feet away from a tree or pole in airtight containers or between trees 100 yards apart. Platforms 15 to 20 feet tall, with metal-wrapped support poles, may also be available.

- If a bear approaches a campsite, people should group together and make as much noise as possible. Children can be sent inside a tent or camper.

- If you hear a bear outside your tent, start talking calmly, turn on lights, and get your bear spray ready. If the bear claws at the sides of the tent or bites at you, treat it as an attack and fight back loudly and aggressively.

When camping in vehicles, follow the practices listed above and all campsite regulations. Keep children close and dogs on leashes. Keep bear spray close and use flashlights after dark. Some campgrounds do not allow food storage in vehicles and require campers to use bear boxes or hang food. Bears can damage or break into cars, trucks, and RVs.

HUNTING AND FISHING GUIDELINES

- Be aware that using elk calls or bugles may attract bears.

- Field-dress game animals immediately, if possible. Move gut pile 100 yards from carcass, at least 100 yards from trails. If you can't pack out a meat carcass, hang properly. Be alert for bears near the carcass at all times. If you return, approach from upwind. Do not defend a carcass from a bear.

- Do not leave fish entrails on the side of a lake or stream — sink them in deep water. Store fish properly.

- Remove clothing used for cooking or cleaning fish or meat.

- Do not burn or bury scraps or waste. Carry out well-wrapped garbage or dispose of it in approved bear-proof containers.

- Use a perimeter alarm system and/or portable electric fence. Portable electric fencing is recommended for hunting camps, long-term camps for livestock or other field workers, or any camp in areas of high bear density.

Bear-Resistant Storage Required

Many federal and state recreation areas require appropriate storage of human and animal food, garbage, cooking utensils, stoves, and scented articles. For a list of certified canisters, panniers, and coolers, see igbconline.org.

Bear Attack: Defending Yourself

A bear that approaches silently, is staring, and is walking on all 4 legs is more likely to be predatory. Immediately respond aggressively and use bear spray. The bear may circle or reappear as you retreat, so remain very alert.

Try to identify the bear as a black or a grizzly, since your response to an aggressive threat will differ. Follow the same protocol for both, however, up to the point that an attack is imminent.

- Don't approach a bear or a cub. Mothers usually send cubs up a tree and then retreat or attempt a bluff. Cubs may moan or wail in fear.

- Don't throw or drop food.

If the bear seems unaware of you, yawning, or not looking at you, do not run but move away quietly and keep watching it. Avoid eye contact. Don't block the bear's escape route. Although bears often stand on hind legs to get a better view or smell, they attack on all fours. Standing up is not a threat.

- Hold bear spray in your hand.

- Keep your pack on as protection.

- Pick up small children so they won't run.

If a bear approaches as you back away, make yourself larger, slowly wave your hands or sticks, and talk in a low voice. Don't say the word "bear" because people who feed bears often say "Here, bear."

- Don't throw things at the bear, but you can drop objects (not food) down or away to distract it.

- Continue to back away. Don't run unless safety is extremely close. At 35 to 40 mph, bears can outrun you uphill or downhill. Don't climb a tree or attempt to swim away. Black bears and smaller grizzly bears can climb trees. Larger bears can ladder up trees, reach high

into trees, or knock you out of a tree. Bears are excellent swimmers that hunt in water.

If the bear continues to approach, making snorts, grunts, chomping, snapping, or deep-throated pulsing sounds, or if it makes mock charges:

- Make loud noises, clap, yell, whistle, blow an air horn, raise your arms, and stare at the bear. If in a group, stand shoulder to shoulder.

- Use bear spray at a distance of 40 to 50 feet away.

- If the bear attacks, continue to use bear spray and fight back aggressively using your hands or objects. Aim for the bear's eyes.

Grizzly Bear Attack

At the point of an actual attack by a grizzly bear, try to judge whether the bear is defensive or predatory. A defensive attack by a grizzly can occur when the bear is surprised, defending a carcass, or protecting her cubs. If your aggressive response and use of bear spray is not effective, you must "play dead" until the bear believes the threat has ended:

- Lie on your stomach with your hands protecting your neck, and brace yourself with your elbows to resist being turned over. Your backpack can offer some protection.

- Draw up your legs, lie still, and remain quiet.

- When the bear leaves, stay immobile as long as possible. If it is nearby, the bear may react again to movement.

If the bear attacks, continue to use bear spray at his eyes, and fight back aggressively using your hands or objects — a fixed-blade camping or hunting knife, a walking stick, a heavy stick, or a rock. Aim for the eyes.

DAMAGE ID: **Grizzly Bear**

PREY ON

Livestock, poultry, beehives

TIME OF DAY

Day

METHOD OF KILL

▸ Distinctive odor, described as musky or musty; claw marks or hair rubbed off on brush or bark; scat evident

▸ Larger livestock run at and slashed from the rear, pulled and held down; head or neck bitten, skull possibly crushed and back or neck broken

▸ Claw marks on animal's face or shoulders from slashing blows or bites delivered while grasped by bear

▸ Small animals bitten through forehead, head, or neck; smallest animals almost entirely eaten, leaving only rumen, skin, and large bones

▸ Carcass held by bear with its feet, underside torn open, heart, liver, and udders eaten; rumen and intestines spread out on ground

▸ Hide sometimes skinned; fat layers may be eaten; a great quantity of the meat eaten.

▸ If killed in open area, carcass may be dragged to a more secluded area; may be covered while bear rests nearby.

▸ Beehives broken and scattered, bear returning until all is eaten

TRACKS

Front 7–13½ inches long, 5–8¾ inches wide; rear 8¼–14 inches long, 4⅝–8½ inches wide. Hind looks like human footprint. Tracks appear pigeon-toed. Claw marks are visible and may extend out from front paw 4 inches. Toes in straighter line than black bear's.

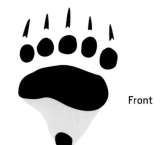

Front

Rear

GAIT

Usually amble or pace at the walk, moving limbs on one side of the body together, stride 25–33 inches. May also direct-register or overstep. Lope or gallop rarer.

SCAT

Blunt or short taper; 7–10 inches long and 1½–2 inches in diameter; left uncovered at site of a kill; black to brown color; can resemble human feces, except contains hair and bones.

7–10"

CHAPTER 6

Weasels

Mustelidae

Weasels, or Mustelidae, belonging to the suborder Caniformia (dog-like mammals), live on all continents except Australia, Antarctica, and many oceanic islands. Excellent hunters, they are variously adapted for digging after burrowing animals, hunting on land, climbing trees, and/or living in water.

OVERVIEW. The members of this family vary tremendously in size from the least weasel, the size of a mouse, to the fierce wolverine. While some have a more substantial physique, many have long bodies with short legs. Slender, quick, and agile, the very active weasels have a high metabolic rate, which requires them to eat about a third of their weight every day.

Mustelids are **bounders** that often travel at a lope in different patterns. In the **rotary** lope, the front and rear feet on the same side of the body often land in the same space. In a **transverse** lope, the rear feet land in the tracks of the front feet. True bounding occurs when the rear feet push off and land at the same time, often overstepping the front feet. Weasels tend to walk only when investigating potential prey or scenting, except for the very large or heavy members of the family.

In North America, mustelids include *Gulo,* the wolverine; *Martes,* the martens and fishers; *Mustela,* the weasels and ermine; *Neovision,* the mink; *Taxidea,* the American badger.

All members of this family have non-retractable claws, often used for digging.

Mustelids have a short nose, or **rostrum**; short, rounded ears; long canine teeth; and **carnassial** teeth adapted for shearing with a powerful bite. Their thick fur is often valuable for pelts — particularly ermine, sable, and mink. Anal scent glands are used for scent-marking territories.

Some are social, while others lead very solitary lives. Many are **polygynous,** mating with different mates during **estrus**, and females may utilize the reproductive strategy of delayed implantation of embryos. The female nurses, feeds, and protects her young in a den until they are about 2 months of age.

HUNTING AND FORAGING. Mustelids are usually carnivorous, whether based on water or land. Vision, hearing, and smell are their key senses for hunting, often nocturnally. Terrestrial mustelids sometimes prey on animals larger than themselves, and occasionally eat plant material. Although they prey on poultry as well, they also provide valuable rodent control. They are, in turn, the prey of larger predators, snakes, or birds of prey.

WOLVERINE
(Gulo gulo luscus)

The elusive wolverine is the largest land-dwelling member of the weasel family in North America. Its reputation as a savage predator is exaggerated, although the wolverine, or "little wolf," is capable of killing animals larger than itself.

Primarily found in the wilderness, wolverines avoid developed areas and are only rarely encountered by humans. Worldwide, their range includes Scandinavia, northern Russia, and Siberia. The North American wolverine is regarded as a distinct subspecies from its Eurasian cousin. The historic North American range extended south from Canada into upper New England, upstate New York, and the Great Lakes states, although possibly not as year-round habitat. In the West, wolverines were found in the Rocky Mountain, Cascade, and Sierra Nevada ranges. Historically, wolverines were hunted both for their fur and to reduce their raiding of trap lines.

At present, wolverines are extremely rare in the United States. They are found in the North Cascade and northern Rocky Mountains, and rarely in the California Sierra Nevada. Recently a wolverine was killed in North Dakota, after an absence of 150 years. Although Michigan is dubbed the Wolverine State, actual wolverines have been absent since the early 19th century, until 2004 to 2010, when a lone female was observed in the "thumb" area of the Lower Peninsula. In Canada, wolverines have not been observed in their historic range in Labrador since the 1950s and Quebec since 1978.

Wildlife biologists are focusing efforts on locating wolverines and studying their behavior to identify their required habitat. Climate change and reduced snow on the mountains will adversely affect the species, which depends on spring snowpack. The need for large ranges also makes preservation difficult, as human development isolates breeding populations.

Surplus Killing

Weasels, along with other predators such as canines, mountain lions, and bears, are often described as bloodthirsty and ruthless when they kill a large number of animals at one time. This reputation stems from the behavior known as **surplus killing**. Surplus killing is a response to an abundance of prey. Livestock and poultry are especially vulnerable when they are gathered together in close quarters. As they attempt to flee, they repeatedly trigger the prey response in the predator. Surplus killing of wild prey is more common in fall and winter, when it is a seasonal strategy to accumulate extra food to cache whenever available.

DESCRIPTION

Stocky, muscular, and bear-like, the wolverine has a thick, sturdy neck and body, with short legs. It has a broad head, small eyes, small and rounded ears, and powerful jaws and teeth capable of rapidly crushing and tearing frozen carcasses. One can easily confuse the wolverine for a small bear, a badger, or a fisher.

Like bears, wolverines walk on the soles of their wide feet and are especially well suited to deep, soft snow. Strong, semi-retractable claws make them agile tree climbers, although they usually stay on the ground. They have poor eyesight and hearing but an excellent sense of smell.

The anal musk glands produce a powerful scent used to mark territory or food sources. These traits are the source of the common name *skunk bear*.

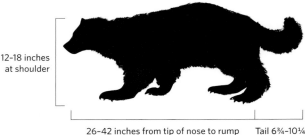

12-18 inches at shoulder

26–42 inches from tip of nose to rump | Tail 6¾–10¼ inches

▲ Wolverine

Size. Male wolverines are significantly larger than females, weighing 24 to 40 pounds or more, measuring 26 to 42 inches long, and standing 12 to 18 inches tall. Females range from 13 to 26 pounds.

Coat and coloration. The coat is dense, soft, oily, and long, and the fur is resistant to frost, which

Although they have a fearsome reputation, wolverines are rarely seen and are not known to attack humans.

was why it was used in the hoods of winter parkas. The color is dark brown to black, with silver markings across the forehead and a creamy white to gold stripe running along both sides from shoulders to tail. Each wolverine has distinctive markings, which allows researchers to identify individuals. The bushy tail is short.

HABITAT AND BEHAVIOR

Wolverines are highly adapted to areas with cool summers and long winters. They are found in arctic tundra, **taiga** (subarctic forest), alpine meadows, and boreal forests. They are not found where normal temperatures rise above 70°F. Female wolverines also need sufficient snow late in spring to dig dens deep in the snowpack. In the southern reaches of their range, wolverines inhabit higher altitudes.

Solitary and elusive, males have large ranges overlapping those of several females. Males in Alaska have ranges from 200 to 260 square miles, which may overlap with females or relatives but not with strange males. Males will defend their territory. Wolverines may construct dens in caves, rocky areas, abandoned burrows, avalanche debris, or fallen timber. Females often use deep snow dens with their offspring.

Wolverines are active all year and have great endurance, traveling up to 40 miles per day in search of food.

Hunting and foraging. *Gulo* means glutton, and wolverines possess the weasel trait of surplus killing. They are very opportunistic consumers of whatever food is available and will cache food. Generally nocturnal, they will also hunt or travel during the day. In winter and spring, they primarily eat the carrion of large animals killed by wolves, other predators, or natural causes. In warmer months they hunt small- to medium-sized mammals, including small predators, newborns of larger animals, eggs, and birds. More rarely, they kill larger animals such as wild sheep, deer, elk, moose, reindeer, or livestock — usually when floundering

in deep snow, because the wolverine cannot outrun faster animals on firm ground. Wolverines will use brush or rocks as cover to ambush prey. Wolverines also climb trees and wait for prey to pass underneath and then pounce or drop onto the back.

Wolverines will forage carcasses found in traps or raid deserted cabins. They fiercely defend their food from lynx, wolves, and bears. They will also consume some plant materials such as berries, seeds, and roots.

▲ The US Fish and Wildlife Service estimates that only 250 to 300 animals are found in the North Cascades and the northern Rocky Mountains, and far fewer in California. Most wolverines live in northern Canada, south into British Columbia and Alberta (estimated population: 15,000 to 20,0000) and Alaska (actual population unknown). Singles or migrants are occasionally seen in parts of their historical range, including lower Michigan, Lake Tahoe in California, Colorado, North Dakota, Utah, and southern Canada.

LIFE CYCLE

Breeding from May to August, females can delay embryo implantation until late fall or early winter. Males mate with 2 or 3 females, breeding with the same mates year after year, and they will sometimes interact with the females and their young kits. Females usually mate only every 2 years, since providing nutrition to raise kits requires a substantial effort.

Litters of 2 or 3 white kits are born February to April or May. The mother may need to leave them alone for several days at a time when searching for food, and she may move the kits if they are disturbed. Although kits become independent at 5 or 6 months, they tend to remain in their mother or father's range until the following spring, often traveling with them. Females often settle nearby while males may travel over thousands of square miles until finding their own range 1 or 3 years later.

Young wolverines are vulnerable to many predatory animals or birds. Upon reaching adulthood, they may survive to age 12 to 13. Adults can fall prey to large predators such as wolves, mountain lions, or bears; other wolverines; starvation; consuming poison baits; or trapping where allowed. When individuals travel farther out of their range, they are susceptible to accidents and increased human interactions.

HUMAN INTERACTION

Preferring to avoid areas used or developed by humans, wolverines are generally found in backcountry and high country. If development pushes into their habitat, interactions with humans may increase. They may damage buildings or food caches in the backcountry in search of food, often spraying musk throughout the area.

Wolverines are efficient scavengers and hunters of smaller prey, only rarely hunting larger animals. Wolverine predation on sheep and reindeer is documented in Scandinavia; however, since wolverines are limited to remote areas, this is not generally a problem in North America. Wolverines in North America also frequently scavenge carcasses killed by larger predators, which are more plentiful than in Europe.

LEGALITIES

To date, the US Fish and Wildlife Service (FWS) has declined to list the wolverine as a threatened species under the Endangered Species Act (ESA). In 2016, federal court action required the FWS to make a listing determination.

Regulated trapping is allowed only in Montana and Alaska. Efforts are under way to extend ESA protection to the wolverine as a threatened species in the United States. It is a species of concern in Canada due to low numbers, low densities, and vulnerable habitat. The eastern Canadian population was listed as Endangered in 2003.

Wolverine Encounters

There are no documented cases of wolverines attacking humans. They are comfortable in the open, at or above the tree line, and will travel on packed roads or trails. If encountered, wolverines may appear curious and may approach you due to their poor vision. Do not confront them, as they can certainly defend themselves. Do not disturb wolverine dens if you encounter them in backcountry, even if it appears that the mother is missing.

Wolverines may be aggressive to dogs accompanying owners in wilderness area recreation. If you have dogs in remote areas, keep the dogs close and inside from dawn to dusk.

Dealing with Wolverines
Livestock Husbandry

In areas where animals graze in wolverine habitat, flocks should not be left alone in spring, early summer, or when vulnerable young animals are present. Untended animals are most vulnerable, as are very young or large, heavy stock that cannot outrun wolverines. Wolverines also seek out weak or floundering animals in deep snow. Pairs or packs of livestock guardian dogs (LGDs) are a useful deterrent.

DAMAGE ID: Wolverine

PREY ON
Livestock

TIME OF DAY
Nocturnal

METHOD OF KILL

▸ Animal killed by a crushing bite to the back of the neck, the back itself, or withers; neck tendons sometimes severed

▸ Less frequently, front of neck or throat bitten. Wolverine will remain clamped on throat of larger animals until they suffocate.

▸ Slashing injuries often present on neck, back, or sides, with deep wounds

▸ Bones sometimes crushed and eaten

▸ Carcasses cached in snow

▸ Carcasses, including those of domestic livestock, often scavenged in winter and spring

TRACK

Front 3⅝–6¼ inches long, 3½–5¼ inches wide; rear 3⅝–6 inches long, 3¼–5¼ inches wide. Claw marks may show; may resemble a large dog; fur may obscure track.

Front

Rear

GAIT

Lope most common with stride of 7–40 inches; feet may strike ground in rotary or transverse pattern.

SCAT

Tapered both ends; 4 inches long, ½ inch in diameter; black to brown; may resemble coyote scat.

4"

FISHER
(Martes pennant)

Found only in North America, the forest-dwelling fisher is a member of the marten family and related to the weasel, wolverine, and badger. It is often called the black cat, the black fox, or the fisher cat. Early colonists saw a similarity to the European polecat, which was also called *fitch* or *fitchet*. In some areas the Native American names *wejack* and *wuchak* are still used, or *pekan* cat, from the Abenaki. Before European settlement, the fisher's range included much of Canada's boreal forest, the Great Lakes states south into the Appalachians, and east through Pennsylvania, New York, and all of New England. In the west, the fisher ranged south into California and the northern Rocky Mountains.

Widespread logging severely reduced fisher habitat in many areas, and with added pressure from fur trappers, the animal disappeared from much of its US range by the early 20th century. Trapping bans in the 1920s and '30s, the recovery of forested land, and deliberate reintroductions have helped the fisher reestablish itself in parts of its historical range. Fishers are more abundant in Canada and the eastern United States, and remain rare in the western areas of their range. They could not be successfully farmed for fur, although it was attempted.

DESCRIPTION

Although related to weasels, the fisher has a larger and less elongated body. Other distinguishing features include pointed, cat-like ears; a broader face; large paws; a long, dense coat and a bushy tail; and semi-retractable claws. The large feet, although not webbed, serve as snowshoes in winter.

Fishers are good climbers, moving quickly through trees, and can descend trees headfirst. They

◄ Valued for their role in maintaining forest and timber health, fishers are being reintroduced in several areas.

are also good swimmers both above and under water. Fishers mark territory with scent from their anal glands and can release it when frightened.

Size. Male fishers range from 37 to 47 inches long and weigh 8 to 13 pounds, with females one-third to one-half the size of the male.

Coat and coloration. The shiny coat is shaded dark brown to black with a white or cream-colored patch on the chest or underparts. The face, head, and shoulders can be grizzled gray or gold.

HABITAT AND BEHAVIOR

Fishers prefer old-growth and coniferous forests but are also found in mixed and softwood large-growth or upland forests and forested wetlands. As their population recovers, the animals are now seen near agricultural areas, suburbs, and occasionally more urban environments, increasing their contact with humans. There is also some evidence that fishers in the eastern United States are increasing in size.

The fisher's requirement for at least 80 percent forest cover and mature trees does limit its movement back into all areas of its former range. The animals will often move through their range in trees and rest on tree branches. They den in hollow logs or trees, abandoned dens of other animals, and rock or brush piles.

Hunting and foraging. Usually solitary, fishers hunt at night, most actively during dawn and dusk, but will hunt in daylight in remote areas. They are aggressive and active hunters, traveling over a large range. They do not hibernate but will cache excess food.

Fishers have a preference for snowshoe hares and are skilled at hunting porcupines — quills are often found in their scat. Primarily hunting on the ground, they also prey on rabbits, rodents, squirrels, small mammals such as raccoons or young bobcats, and ground birds. Fishers will at times scavenge carcasses and eat insects, amphibians, nuts, and berries, but not fish. On farms, fishers will take poultry or small mammals such as rabbits. They are widely reputed to prey on domestic cats and small dogs, but studies of scat and stomach contents of trapped fishers in New Hampshire, New York, and Massachusetts do not confirm this.

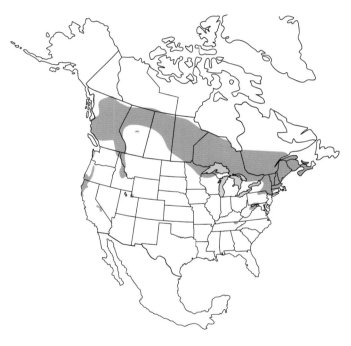

▲ Today the fisher is primarily found across much of the Pacific Northwest from British Columbia, northern Alberta, and Saskatchewan to Manitoba, Ontario, Quebec, New Brunswick, and Nova Scotia; portions of Montana, Idaho, Minnesota, Wisconsin, Michigan, New York, Pennsylvania, and West Virginia; much of New England; and pockets in Oregon, northern California, Utah, and Tennessee. Fisher reintroductions in the south and central Cascade Mountains in Washington began in late 2015.

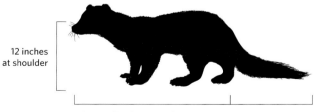

12 inches at shoulder

20–31 inches from tip of nose to rump   Tail 12–16 inches

▲ Fisher

Evading the Quills

One of the few predators to take on a porcupine, the much small fisher rapidly circles around the larger animal and repeatedly bites at its exposed face while evading its back, tail, and quills. The battle can continue for a half-hour or more until the porcupine is weakened and the fisher can move in to flip it and attack the vulnerable belly. At other times, the fisher climbs up a tree after a porcupine, and then maneuvers to attack it from above.

LIFE CYCLE

Fishers reportedly breed with several mates, although male and female pairs have been seen together and with young offspring. They mate in the summer after the female bears her young, but implantation does not occur until the following January or February. A month later, the female gives birth to 1 to 4 kits in her den, usually located high in a cavity in a large tree. When the kits become mobile the parents move them to a ground den. Kits grow rapidly and reach adult size at 3 to 4 months of age, soon becoming independent and breeding 1 or 2 years later.

Fisher life span can be as long as 10 years in the wild. Larger predators or birds of prey may prey upon young fishers, but it is rare for an adult to be killed. Trapping, where allowed, also accounts for their mortality.

HUMAN INTERACTION

Fishers will prey on poultry and other game birds, as well as domestic rabbits. In the northeastern areas, fishers are increasingly being seen during the day as they adapt to human presence in the suburbs and exurbs.

The loss of fishers is linked to an increase in the number of porcupines, which can heavily damage small and large trees. Fishers are one of the few known porcupine predators.

LEGALITIES

Fishers are a protected species in some states, with regulated trapping seasons elsewhere. Consult with state wildlife authorities.

Fisher Encounters

It is rare to encounter a fisher. If threatened or cornered, a fisher may attack humans or dogs. There are a few incidences of attacks on children but no recorded human fatalities.

Dealing with Fishers
Livestock Husbandry

- Reduce rodent populations. Bird feeders and spilled feed attract squirrels and rodents, which in turn attract fishers.

- Safe containment is essential for preventing attacks on poultry or rabbits. Fishers will chew through wood or soft wire. Cover all small openings and gaps with hardware cloth or welded wire.

- Cover the top of the pen as well. A hardware cloth apron will prevent digging.

- You may need electric fencing to discourage a determined fisher.

DAMAGE ID: Fisher

Note: Fishers are instinctively triggered to kill multiple birds at one time, a practice known as **surplus killing**.

PREY ON

Poultry and rabbits

METHOD OF KILL

▸ Animal killed by bite to back of head or neck, sometimes severing it. If first bite is elsewhere on body, fisher will grasp prey tightly and relocate bite to back of neck.

▸ Claw marks left on body

▸ Chicken or rabbit carried away

▸ Chicks or young rabbits eaten whole

▸ On a carcass, head and neck eaten; breast or rear end sometimes bitten open and intestines pulled out

TRACK

Front 2⅛–3⅞ inches long, 1⅛–4½ inches wide; rear 2–3⅛ inches long, 1½–3½ inches wide. Small claw-tip marks often visible. Can be confused with a marten; lacks webbed feet of a mink or otter. Tail or body impressions may be seen in snow.

 Front

 Rear

TIME OF DAY

Nocturnal, most active at dawn and dusk

GAIT

Primarily rotary lope, stride 6–30 inches. Mat walk or transverse lope in deep snow.

SCAT

Long, slender, tapered at ends; 3½ inches long and ½ inch in diameter. Often left on logs or high sites; may contain porcupine quills.

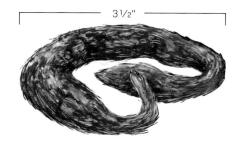

3½"

AMERICAN BADGER
(*Taxidea taxus*)

Related to the Eurasian and honey badgers, the American badger was named for its resemblance to the common badger found in Europe. It was not present in the original eastern colonies, and Lewis and Clark and other naturalists described it as they encountered it. The historical range included grassland or prairie in south-central Canada and the United States, northwest to southern British Columbia, and south to Mexico. Widely hunted for food and pelts by both Native Americans and European settlers, it expanded its range east and north as land was cleared for agriculture and was first sighted in Ohio in the late 1800s, and more recently in Ontario. The badger is now less common due to habitat and prey loss, especially in the North. It is seen most often on the Great Plains.

DESCRIPTION

Size. The American badger weighs 14 to 30 pounds and ranges from 20 to 26 or more inches long, with the male considerably larger than the female in all four subspecies. A powerful and stocky animal, the

▲ If confronted, badgers will hiss or snarl ferociously, make short charges, and bite if cornered, but they usually retreat from most threats into a den.

badger has a broad head; a sturdy neck; a flattened, low body shape; thick skin; and short but large, strong legs. The front legs are equipped with long claws for powerful digging, while the hind claws are short and shovel-like. The tail is short and bushy.

Coat and coloration. The badger has a grizzled silver-gray to yellow-brown coat with white to buff underparts and black or dark brown feet. Distinctive head markings include a white stripe over the top of the head and brown or black triangular patches on each cheek.

HABITAT AND BEHAVIOR

Badgers prefer the open countryside of grasslands, high meadows, and desert scrub or grasslands. They prefer soft soils, which allow them to dig for prey and dig their burrows.

American badgers are solitary, moving frequently in search of food. They dig temporary homes with a single tunnel and a shallow **sett**, or den, or they enlarge the dens of their prey, unlike the European badger, which inhabits deep and permanent warrens. The elliptical entrance is 8 to 12 inches in diameter and usually has a large mound of fresh dirt piled in front. Females dig deeper and larger setts for raising young, but they may move their cubs to new dens if needed.

Although a badger will enter a torpor state in its den during very cold weather, it does not truly hibernate. Males utilize larger home ranges than females and may mate with more than one female in overlapping ranges.

Hunting and foraging. Badgers prey on burrowing rodents (ground squirrels, pocket gophers, prairie dogs, moles, voles, rats, and mice) by rapidly digging. They also hunt ground-nesting birds and eggs, as well as small mammals, snakes, lizards, and amphibians. Coyotes will sometimes opportunistically hunt near badgers as they dig. Badgers will also eat insects, carrion, and some garden vegetables. Primarily nocturnal, badgers are active during dawn and dusk. Females with cubs may forage during the day.

LIFE CYCLE
Although badgers mate in summer or fall, embryo implantation is delayed and the 2 to 5 cubs are not born until the following spring. Cubs remain with their mother until midsummer, and females may bear young the following year. Wild badgers may live until 14 years of age.

Only large predators will hunt an adult, while young badgers are vulnerable to birds of prey and coyotes. Others die from hunting, trapping, poisoning by rodenticides, or road accidents.

HUMAN INTERACTION
Badgers are a valuable predator of rodents, rattlesnakes, and pest species.

Their digging does create dangerous holes in pastures and agricultural fields, occasionally damaging irrigation ditches or dams and the sides of roads. Badgers will also eat ground-nesting birds, young poultry (including turkeys), domestic rabbits, and eggs. They have been known to kill small lambs. They may also bother beehives in search of bees or honey and occasionally disturb gardens in search of corn, peas, or green beans.

LEGALITIES
Badgers are listed as an Endangered species in Ontario (which has about 200 badgers) and British Columbia. They are also protected in Arkansas, California, Illinois, Michigan, and Wisconsin (the Badger State). Some states have regulated trapping seasons. Consult authorities before removal.

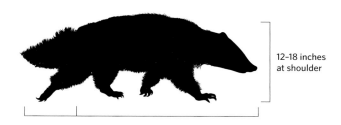

12-18 inches at shoulder

Tail 4-6½ inches   20-26+ inches from tip of nose to rump

▲ American Badger

▲ Badgers are found from the Great Lakes states westward through the Great Plains to California, and south to Texas. In Canada, badgers are found from southern British Columbia, across the plains into southwestern Ontario.

Badger Encounters

Although people do not generally encounter badgers, the animals will react defensively when threatened. If a badger surprises you, do not confront it. Avert your eyes and back slowly away until you are about 100 feet off. At that distance, you can safely observe it.

Dealing with Badgers
Farm Buildings, Homes, and Yards

The best prevention is the control of rodent populations around buildings and fields. To protect pens and coops, bury mesh fencing 12 to 18 inches deep with outward aprons of 18 to 24 inches. High-intensity floodlights tend to discourage badgers near buildings and yards.

DAMAGE ID: Badger

The most common sign that a badger has visited is evidence of powerful digging, with numerous holes dug in search of prey.

PREY ON

Chickens, turkeys, ground-nesting game birds, rabbits, small lambs or kids, beehives

TIME OF DAY

Nocturnal

METHOD OF KILL

▸ Fences dug under; pens or coops dug up through dirt floor

▸ Poultry, rabbits, or lambs usually completely consumed except for head and fur or feathers

▸ Musky scent left behind

▸ Body parts of larger animals buried nearby

GAIT

Usually walk; stride 5–10 inches; direct-register. Will travel at a trot, with the same stride, and rarely lope.

SCAT

Blunt, 3 inches long and ¾ inch in diameter; usually left underground

TRACK

Front 2⅞–3⅞ inches long, 1½–2⅝ inches wide; rear 1⅞–2¾ inches long, 1⅜–2 inches wide. Long claw marks usually visible.

Front

Rear

WEASELS
(Mustela)

The weasel certainly has a poor reputation, and the word itself has become synonymous with deceit and treachery. At the same time, the pelts of one weasel, the ermine, appeared in Native American ceremonial clothing and headdresses, as with the native ermine in Europe and Asia, where the white fur with black-tipped tail spots adorned the cloaks of royalty.

Of the larger weasel family found nearly worldwide, 4 species live in North America, including the long-tailed weasel, the short-tailed weasel or ermine, the least weasel, and the endangered black-footed ferret. The white winter coats of both the ermine and the long-tailed weasel were sold as ermine in the fur industry, although demand for these pelts has now lessened.

Although their natural ranges vary, the weasels are much alike in appearance and behavior. They have excellent senses of smell, hearing, vision, and touch. Their quickness and agility are essential not only for hunting prey but also for escaping predators. Fierce and bold, a weasel will attack larger animals with its powerful bite.

DESCRIPTION

Highly adapted to pursuing prey in tunnels or burrows, weasels are long-bodied and slender, with short legs; pointed faces; and small, rounded ears. Males are larger than females, and both sexes can produce a strong musk from their anal glands. With its short front legs and its back arched up, the weasel moves with a bounding gallop, the rear feet almost landing in the impressions of the front feet.

HABITAT AND BEHAVIOR

Home range size and population density vary among the weasel species and also depend on the availability of food, the season, and other factors. Males have larger territories, but both sexes defend their ranges. Although long-tailed weasels can dig their own dens, many weasels use abandoned burrows of other animals, hollow logs or trees, brushy or grassy patches, rock or debris piles, or even buildings.

Hunting and foraging. Hunting in trees, on the ground, in the underground burrows of rodents, across snow or under it, weasels are very active with high-energy demands to eat frequently. They primarily prey on small rodents, including pocket gophers, muskrats, and young rabbits, and to a lesser extent on snakes and birds to feed on their blood and their eggs. They will also eat invertebrates, amphibians, and small fish. They may eat fruits during summer.

Weasels attack with great speed, grasping their prey by the back of the head or neck with a powerful bite, wrapping their body around the victim, and holding it with their feet. **Surplus killing** behavior is common in weasels, and they will cache carcasses for future eating. This instinctual behavior has contributed to their reputation as bloodthirsty killers.

Weasels are solitary animals, but are often comfortable out in the open. Although commonly believed to be nocturnal, in fact they are often active throughout the day and night, taking frequent rests back in their den. Weasels do not hibernate in winter, emerging to hunt or feed on cached food.

LIFE CYCLE

Mating polygamously in the summer months, males will fight over females. Females are delayed ovulators, bearing their litters of 4 to 9 kits in the spring. Fully grown at 3 months, females are capable of mating that summer. Weasels do not hibernate. Although they may have a life span of 3 to 7 years, many do not survive even 1 or 2 years. They are the prey of larger weasels, larger predators, rattlesnakes, and birds of prey.

HUMAN INTERACTION

Weasels are very beneficial predators of rats, mice, and other rodents such as pocket gophers, which damage grazing plants and crops such as alfalfa. They may kill poultry or feed on eggs. The long-tailed weasel will also prey on rabbits.

Weasel Encounters

If seen, a weasel will quickly disappear. Weasels have attacked humans who attempt to take their prey away or grab them. If threatened, they can be fiercely aggressive to larger animals such as cats or dogs.

LEGALITIES

As furbearing animals, weasels can be hunted in most states during specific trapping seasons. Consult local and state authorities before removing.

Dealing with Weasels

- Reduce the rodent population to reduce weasels' interest in outbuildings.

- Note that many frightening techniques are not effective with weasels, although motion-activated sprinklers may discourage them. Dogs, especially LGDs, can also be helpful.

- Exclude weasels from coops and rabbit pens or cages to prevent weasel damage to poultry.

- Since weasels can slip through a 1-inch opening, cover all gaps with wood, metal, hardware cloth, or mesh with no larger than ½-inch square openings.

Long-Tailed Weasel
(*Mustela frenata*)

The most common weasel in North America, the long-tailed weasel may appear around farms or suburban areas, leaping into roadside ditches or hedgerows near fields or pastures. A good climber and swimmer, the long-tailed weasel is adaptable to habitats ranging from alpine to tropical, as long as there is water for drinking and cover for hiding. In Canada and the United States, it prefers open woods, and brushy, grassy, or marsh-like areas. Usually nocturnal, the long-tailed weasel can switch to **diurnal**, or daytime, activity in summer.

Size. The long-tailed weasel ranges from 13 to 24 inches and weighs 12 to 16 ounces. It earned its name from its tail, which is more than half of the combined heady and body length.

Coat and coloration. Appearing in different colorations across its range, the long-tailed weasel is found in shades of dark to medium brown, reddish brown, or orange-brown on the body, tail, and feet. The toes may be tipped in white, the throat and underparts may be white or orange, and the furry tail has a black tip. In northern areas, the coat color changes to white or yellowish white in winter. The dark-masked, orange-brown variety found in Kansas, Oklahoma, and Texas is often called the bridled weasel.

▲ The long-tailed weasel prefers fresh prey and will move his head from side to side when hunting, seeking out telltale scents.

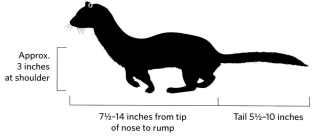

Approx. 3 inches at shoulder

7½–14 inches from tip of nose to rump

Tail 5½–10 inches

▲ Long-Tailed Weasel

◀ Long-tailed weasels are found throughout the United States except the very arid southwest, into Central America and northern areas of South America. In Canada they range from southern British Columbia, across the prairie provinces to southern Ontario, Quebec, and New Brunswick.

Short-Tailed Weasel or Ermine
(*Mustela ermine*)

This weasel is also called the stoat in Europe, where it was long valued for its winter fur pelt, which is white. The ermine is primarily found in Canada, Alaska, the northern states, and parts of the West. Common in its range, it survives well in open coniferous and mixed woodlands, marshes, tundra, brush, open ground, and agricultural fields. Also good climbers and swimmers, ermines are active both day and night.

Size. With a shorter tail than its cousin, the ermine ranges from 7 to 13 inches long and weighs 1.5 to 6 ounces.

Coat and coloration. The ermine has a reddish to dark brown back and tail, a yellow to white belly, brown feet with white toes or entirely white feet, and a black tip on its tail. In winter the coat turns white, shading to yellow, retaining only the black tail tip.

▲ The dense and silky pelt of the ermine has long been valued as a luxury fur. The desirable white color occurs in winter in the northern portions of its range.

▲ The ermine is found in Alaska, all Canadian provinces, in the upper Midwest through Pennsylvania and north to New England, the Pacific Northwest, and the Rocky Mountain West.

Approx. 2 inches at shoulder

Tail 2–5 inches   5–8 inches from tip of nose to rump

▲ Ermine

Least Weasel
(*Mustela nivalis*)

The least weasel has a northern range similar to that of the ermine but is not found in the western states. Considered the smallest carnivore in the world, this weasel is 5 to 8 inches long and weighs only 1 to 2.5 ounces. Because it is so small it can be mistaken for a baby weasel. Its upper parts are brown, although the short tail lacks a black tip. It has white underparts and brown or white feet. In northern areas, the least weasel is white in winter.

The least weasel is active day and night, hunting with great energy in burrows or under leaf litter on the ground. Its prey includes small rodents, shrews, and moles, and on the tundra it preys on lemmings. Less commonly seen than the other weasels, the least weasel is also found in meadows, marshes, agricultural fields, open woods, or brushy areas.

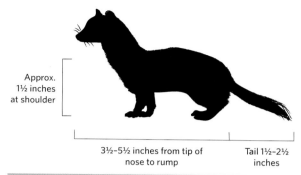

Approx. 1½ inches at shoulder

3½–5½ inches from tip of nose to rump

Tail 1½–2½ inches

▲ Least Weasel

▲ The least weasel's range includes Alaska; the upper Midwest south to West Virginia and into North Carolina and Tennessee; and everywhere in Canada except arctic areas, the Maritime provinces, southern Ontario and Quebec, and central and coastal British Columbia.

Newborn least weasels are about an inch tall and weigh less than one-fifth of an ounce.

DAMAGE ID: Weasel

PREY ON

Poultry, eggs, domestic rabbits

TIME OF DAY

Day and night

METHOD OF KILL

- May surplus-kill multiple birds or rabbits.

- Bloody body and feathers left behind in a pile; piled and cached surplus carcasses

- Head, upper neck, or jugular bitten through for the kill, leaving closely spaced marks of canine teeth; larger birds sometimes bitten several times

- Head alone may be eaten, or back of neck or a leg chewed on

- Eggs eaten, leaving shell with a hole of ½–¾ inch broken in end, with finely chewed edges.

- Weasels wrap their body around the victim and hold it with their feet. They will eat blood, leaving small wounds in neck or body.

TRACKS

Long-tailed: Front ⅝–1½ inches long, ¾–1¼ inches wide; rear ¾–1½ inches long, ½–1 inch wide. May include drag mark of the tail; claw marks may be visible. Ermine track slightly smaller, least weasel even smaller.

 Front

 Rear

GAIT

Long-tailed: Rotary lope, stride 10–45 inches; bounding stride 15–25 inches

Ermine: Rotary lope, stride 4–40 inches; bounding stride 14–22 inches

Least weasel: Rotary or transverse lope, stride 6–38 inches; bounding stride 11–26 inches. Short and long strides can be mixed together.

SCAT

Long, slender, and tapered ends, to 1½ inches long and ¼ inch in diameter; often left on logs or rocks

Least Weasel

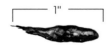

Ermine

Long-Tailed Weasel

MINK
(*Neovison vison*)

Weasels, otters, mink, and ferrets look much alike to the casual observer. The American mink is also hard to distinguish from the European mink, although it is actually more closely related to the Siberian weasel. All European minks have a white patch on their upper lip, while American minks may or may not have that white spot. In general, American minks are larger than their European cousins, with skeletal differences as well. In Canada and the United States, there are 15 recognized geographical subspecies, varying in size and color.

Both Native Americans and European settlers trapped and hunted the American mink for its luxurious pelt. Minks were also partially tamed and used for ratting, in the same way as a ferret. Since the late 19th century, both US and Canadian farmers have raised American minks as fur animals, producing several million pelts annually. Selective breeding of farmed minks has produced several exotic colors and shades, including pure white and black. American mink fur is denser and closer fitting, as well as longer and softer, than European mink. Exported elsewhere for farming, American minks have unfortunately been released or have escaped in many parts of the world, including much of Europe and Russia, where they are viewed as an invasive species and a threat to native mink.

DESCRIPTION

American minks weigh 1 to 3 pounds and measure 18 to 24 inches long, including the tail, with the male larger than female. Larger and more substantial than a weasel, the American mink is long and streamlined to aid it in swimming and entering tight burrows. It has a pointed skull, short ears, a long neck, a slender body, and short legs with webbed feet. The anal glands are used for scent marking and can discharge foul musk about 1 foot. The mink's eyesight and hearing are superior to its sense of smell.

Coat and coloration. The wild mink usually has a rich, shiny, dark brown double coat. White spots may be present on the chin, throat, and chest. The oily guard hairs make the coat water-resistant. The heavier and thicker winter fur is the most desirable.

HABITAT AND BEHAVIOR

Primarily dwelling and hunting on grassy or brushy shorelines, the semiaquatic mink needs a permanent water source and may be found near large lakes and small ponds, streams and rivers, marshes and swamps.

▲ The wild mink is usually found near water.

Approx. 5 inches at shoulder

12–16 inches from tip of nose to rump | Tail 6–8 inches

▲ Mink

It is an athlete: an excellent swimmer, both on the surface and underwater; a diver; a tree climber able to descend headfirst; and a rapid traveler on land, with a bounding stride of 11 to 26 inches.

Though minks prefer forested areas, they inhabit brushy or rocky areas as well. They utilize several burrows in their area, such as natural holes and crevices, abandoned dens, and drains and rocky areas; more rarely, they dig their own homes.

Hunting and foraging. Strictly carnivorous, minks eat fish, crustaceans, amphibians, birds, eggs, insects, and small mammals such as rodents or rabbits. They do not ambush or stalk their prey but rush it, pursuing smaller creatures in their own tunnels or burrows, in water, and on land. Minks are primarily nocturnal but can be seen during daylight. Although they do not hibernate, extreme winter weather may keep them in their burrow for a day or two.

LIFE CYCLE

Minks are solitary by nature, but the home range of a male may overlap those of females in the area. Both sexes are polygamous, and males fight for mates during the late-winter breeding season. The females have a delayed embryo implantation with 3 to 6 kits born later in the spring. Kits stay with their mother until late summer or early fall and are sexually mature at less than 1 year of age.

Most mink fall prey to human trapping or traffic accidents, while natural predators include foxes, bobcats, and large birds of prey. The mink's life span is about 3 years in the wild but longer in captivity.

▲ American minks are found throughout the United States and Canada, except in much of the Southwest and some of the very far North.

HUMAN INTERACTION

Minks are useful predators of muskrats and other rodents. Although farmed for pelts and mink oil, captive animals are not selected for tameness and make poor pets.

LEGALITIES

Most states and provinces have legal trapping seasons for mink. Homeowners should contact state wildlife authorities before trapping or taking other measures.

Mink Encounters

Minks will hiss, snarl, and growl if they feel threatened. They are fierce if confronted by cats, dogs, or other predators; if handled, they are likely to bite aggressively.

Dealing with Mink

- Prevent mink damage to poultry or rabbits with exclusion methods. Since minks can use a 1-inch opening, cover all gaps with wood, metal, hardware cloth, or mesh with no larger than 1-inch square openings.

- Increase brush and tall grasses in waterfowl or game birds' area to reduce mink predation. Note, however, that this will likely increase your animals' vulnerability to other predators.

- Frightening techniques are not effective with minks, although LGDs and other dogs can be helpful.

DAMAGE ID: Mink

PREY ON

Poultry, waterfowl, game birds, rabbits

TIME OF DAY

Night

METHOD OF KILL

▶ Single bird or rabbit killed most often by bite through head, upper neck, or jugular vein, leaving closely spaced marks of canine teeth

▶ Larger birds killed by drowning

▶ May surplus-kill in a small area and pile up the bodies.

▶ Bloody body, feathers, fur, and strong musky smell, similar to skunk, left behind

▶ Head, back of neck, or a leg eaten

▶ Blood eaten, leaving small wounds on neck or body

▶ Eggs eaten, leaving the shells

▶ Minks wrap their body around their victim and hold it with their feet.

▶ Minks will not venture far from water.

▶ Minks do not gnaw through wood but will use openings left by other predators.

TRACK

Front 1⅛–1⅞ inches long, ⅞–1¾ inches wide; rear ¾–1¾ inches long, 1–1⅝ inches wide. More evidence of webbing between toes than with weasels; claw marks may be visible. Snow slide trails may be seen in winter.

 Front

 Rear

GAIT

Bounding stride 11–26 inches; rotary and transverse lope, 6–38 inches; will walk to explore.

SCAT

Long, slender, tapered ends; 2 inches long and ¼ inch wide; black–brown color; may be oily and smell like or contain fish; deposited on logs near water

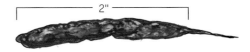
2"

AMERICAN MARTEN
(*Martes americana*)

Related to the fisher, the American marten is also known as the pine marten or American sable. Though more abundant than the fisher, it is restricted to mature areas of coniferous or mixed forests and deeper snow, especially in northern Canada and Alaska and the western mountains. As with the fisher's, the marten's historical range was reduced and reintroductions have occurred in the upper Great Lakes and northeastern states. Populations are threatened or protected in some areas, while open to trapping seasons in others.

DESCRIPTION

Although it resembles a fisher, the American marten is closer to the weasel or mink in size. The males weigh 1 to 3 pounds and range from 18 to 26 inches long, and the females are about one-third smaller. The silky coat ranges in color from pale buff or gray to brown to black, with a lighter head. There is often a pale yellow to orange patch on the throat and chest.

HABITAT AND BEHAVIOR

Quick, agile tree climbers and swimmers, American martens live in remote regions and therefore pose little threat to poultry or domestic rabbits except in specific areas. Prevention is similar to that for weasels.

American martens may be active during the night or day, depending on availability of prey, cold weather, or the presence of young. Martens can be more active in the day during the summer as well. Well adapted to deep snow, they often prey on snowshoe hares in winter. Voles also form a significant portion of the marten's diet, along with red squirrels, meadow mice, or shrews.

Their smaller size makes them more vulnerable to predators. The American marten has a desirable pelt for trappers, with deaths also occurring due to accidents and disease.

◄ American martens are occasionally seen in Maine, upstate New York, Michigan, Wisconsin, and Minnesota. There are ongoing reintroduction efforts in Ontario and Minnesota.

Raccoons

Procyonidae

Found only in the Western Hemisphere, the Procyonidae family includes raccoons, kinkajous, ringtails, coatis, and the less well-known Central and South American olingos, olinguitos, and cacomistles.

Although they may be descended from the Canine family, the members of this family are omnivores. Small in size with long tails, they usually have slender bodies, the raccoon being the exception. Many members have markings on their faces and banded coloration on their tails. They walk **plantigrade**, flat on the ground like bears, with nonretractable claws. Generally nocturnal, they lead solitary lives, with the females raising their young alone.

Now found in 18 species, raccoons are adapted to various habitats from deserts to woodlands, wetlands, and tropical rainforests, with the greatest diversity occurring in Central America. The red panda, once considered a member of this family, is now classified with Ursidae, the bear family. Recent genetic studies have shown that *Procycon*, the raccoon, is most closely related to the ringtails and cacomistles. The Northern raccoon has been found in the United States for 2.5 million years.

NORTHERN RACCOON
(Procyon lotor)

The intelligent, black-masked raccoon has long been famed for its ability to open latches and locks, solve complex situations, and remember its solutions over time. Native American folktales often describe it as a trickster, capable of outsmarting other animals. The Powhatan word, recorded as *aroughcun,* was noted in the earliest days of the Virginia colonies. The Aztecs called it *mapachitli,* or "one who takes everything in its hands," which became *mapache* in Spanish.

The clever raccoon has expanded its range across the country, adapting to life in backyards and cities.

Columbus was the first European observer of the raccoon, originally thought to be a member of the bear family. Early explorers did not find raccoons in the central and north-central areas of the United States, suggesting that the species was limited to the wooded river lands of the southeastern states and south into Central America.

Colonists soon adopted the traditional use of the ringed tail as a hat decoration, and they widely hunted raccoons for fur and meat, often with coonhounds bred to the task. By the 1930s, the population was severely reduced; however, the species has rebounded into the millions and greatly expanded its range.

Related to the Cozumel and crab-eating raccoon species, the Northern raccoon is commonly found from southern Canada throughout the United States, in both urban and rural areas except high mountainous elevations or very arid environments. The success of the species is linked to the loss of natural predators and their adaption to both urban and agricultural habitats. Due to escapes and deliberate releases, raccoons now inhabit some European and Asian countries, including Germany, France, Spain, the former Soviet Union, and Japan.

DESCRIPTION

Raccoons have poor long-distance vision but see well in the dark, and they have good senses of smell and hearing. Their hind legs are longer than their front legs, giving them a hunched stance and limiting their speed and jumping ability, although they are excellent climbers and swimmers and can descend trees headfirst. Their front paws are very dexterous with a highly developed sense of touch.

Size. Today raccoons include more than 20 subspecies that vary in size and coloration, ranging from 24 to 44 inches long, including the long, bushy tail, and weighing from 4 to 40 pounds, although 15 to 30 pounds is more common. Larger and heavier raccoons have been recorded. The largest animals are in the Northwest, and the smallest in the Southeast. Males are heavier than females,

Do Raccoons Wash Their Food?

The word *lotor* in the species name describes the raccoon as a "washer," possibly because it closely examines potential food with its paws, sometimes rubbing at it. Contrary to popular belief, although raccoons may also dip the food in water if available, this action is not for cleaning but an instinctive manipulative behavior possibly related to searching for food on the banks of water.

9–12 inches at shoulder

16–28 inches from tip of nose to rump

Tail 8–16 inches

▲ Northern Raccoon

and raccoons can gain twice their normal weight in preparation for cold winters.

Coat and coloration. The characteristic black mask and ringed tail can appear on a base coat that is grizzled gray, reddish brown, or yellowish. The nose is relatively short, but the skull is broad, with rounded ears.

HABITAT AND BEHAVIOR

Raccoons traditionally lived in hardwood forests, often foraging at the water's edge, but they are highly adaptable to varying climates and terrains and will use livestock troughs and ornamental ponds as a water source. Preferring tree hollows aboveground for their dens and resting spots, raccoons will also occupy the burrows of other animals, caves, brushy or rocky areas, and human structures from buildings to culverts and sewers. Individual raccoons will also move among several dens or resting sites.

Related females often share common feeding or resting areas within their home range, although mothers with kits will be protectively solitary. Males also band together during breeding season to defend against strange males. Range size varies according to availability of food and water. Populations are denser in urban and suburban locations.

Hunting and foraging. Its diet varying according to location, the raccoon is an opportunistic omnivore, eating insects, amphibians, reptiles, fish, crustaceans, small mammals, eggs, scavenged carcasses and garbage, fruits, nuts, seeds, roots, and crops such as corn. Raccoons use their front paws to catch or handle their food. Raccoons are usually nocturnal, although mothers with young kits may forage during the day and raccoons near coastal areas may take advantage of low tides.

▲ The Northern raccoon is found across the southern areas of Canada from the Maritimes to Alberta and coastal areas of British Columbia, throughout most of the United States, and south through Central America into northern South America.

LIFE CYCLE

Breeding season may occur from January to June, depending on the area. A female will often mate with more than one male, but she raises her 3 to 7 kits alone. The kits often stay with their mother for their first winter, although young females are capable of breeding during their first spring. During very cold or snowy times, solitary raccoons or small groups will sleep or be inactive in a den, but they do not hibernate. While raccoons can live up to 20 years, most survive only 1 or 2 years, falling prey to traffic accidents, hunting and trapping, larger predatory mammals and birds, snakes, starvation, and illness. Distemper is especially prevalent in the raccoon population and the leading cause of natural death.

HUMAN INTERACTION

Raccoons can move into homes or other buildings, damaging siding and roofing; ruining fruit and other crops; and killing poultry, waterfowl, game birds, and rabbits. They are among the primary "problem animals" removed by wildlife control professionals.

Raccoons can transmit rabies, plague, canine distemper, or leptospirosis to human or domestic animals. *Baylisascaris procyonis*, a specific roundworm, is a potentially fatal parasite transmitted to humans via feces. Feces should be burned or buried, and surfaces treated with boiling water or high temperatures. Handle with caution and do not inhale particles.

Raccoons are the primary carriers of rabies in the eastern United States, and every state in the country has identified rabid raccoons. Rabies has also returned in the raccoon population in New Brunswick, Quebec, and Ontario. Rabid raccoons will usually, but not always, appear ill, with unusual movements, vocalizations, or aggressive behavior. Raccoons with distemper also display unusual movements and disorientation. Use caution when encountering raccoons during the day, since this may be a sign of disease. The best advice is to leave raccoons alone and handle dead animals only with protection.

The ring-tailed cat (left), found throughout the southwestern United States, is smaller, far more reclusive, and more timid than its cousin the raccoon. The coati, found in a much smaller range in the same area of the country, is the only member of this family that is active in the daytime.

LEGALITIES

Northern raccoons are subject to regulated hunting or trapping seasons in many areas. Check with state wildlife agencies for other regulations. Relocation of trapped raccoons is prohibited in many areas due to the threat of disease.

Raccoon Encounters

Do not attempt to make a pet out of a baby raccoon because it can become very difficult to control as an adult and will be too friendly toward humans to be released. Disease is also common among pet raccoons.

If a raccoon surprises or confronts you, make yourself larger by waving your arms and shouting or throwing stones or water near the animal to chase it off. Teach children to move away and shout, "Go away, raccoon!" to alert adults that they need help. If the raccoon behaves aggressively, do not attack or trap it; instead, call local wildlife authorities. Raccoons rarely attack cats or dogs but may fight if cornered.

Rabies is always a serious concern in any incident with a raccoon.

Dealing with Raccoons
Homes and Yards

- Do not feed raccoons, which encourages their dependency and habituates them to humans, leading to more aggressive behavior.

- Keep outdoor cooking areas clean. If the area is contaminated by raccoon feces, carefully dispose of the feces and disinfect the area and utensils.

- Secure garbage and wild bird, pet, and animal feed.

- Do not feed pets outside or leave pet food outside at night. In more arid areas, do not leave water out at night.

- Secure pet doors at night or use electronic openers. Raccoons have entered homes through these doors.

- Use care when allowing pets out at night. Do not leave smaller dogs outdoors at night except in a protective kennel with a top.

- Wrap ripening sweet corn to the stalk with a double loop of filament tape, which also prevents husks from being pulled back.

- Use safe, commercially designed caps on chimneys and block access to attics, crawl spaces, and under-porch areas. A raccoon can enter through a 4-inch-diameter hole.

- Use metal or plastic spikes, vent piping, or sheet metal to prevent raccoons from climbing up poles, trees, or the sides of buildings.

- Construct fishponds with steep sides. Two-foot-tall mesh fencing or electric wire spaced at 6 and 12 inches will also discourage raccoons. Cover smaller ponds with mesh or netting, especially at night.

Livestock Husbandry

- Cover openings that might provide shelter, and clear areas of rubbish, brush, and overhanging tree branches near animal enclosures and buildings.

- Hardware cloth is the best exclusion device for shelters and pens. Tops need to be covered as well.

- Since raccoons can open many simple latches — such as hook and eye, spring clips, snap hooks, and sliding or lifting latches — use Kiwi latches or padlocks on coop or pen doors and gates.

- Lights, sound, scarecrows, and streamers may provide temporary protection, but raccoons rapidly habituate to scare devices or floodlights.

- Livestock guardian dogs (LGDs) provide reliable protection against raccoons.

Fencing

- Improve fences with scare wires at bottom (at 6 and 12 inches) and top.

- Use temporary electric fencing around gardens and animal areas.

Evicting a Trapped Raccoon

Raccoons are most active beginning at dusk, when eviction methods are most successful. You can construct a one-way door out of a space where a raccoon is trapped; however, very young kits may not be mobile enough to use one. You do not want the raccoon to die entrapped.

If a raccoon falls in a Dumpster or a similar pit, insert a branch or rough board to help it climb out.

If a raccoon enters your house or another building, close interior doors to isolate it in one room. Open the windows and exterior doors in the room, including any pet doors. Do not corner the raccoon, which may defend itself.

Raccoons in chimneys are usually mothers with kits, which are very noisy. Keep the damper closed, bang the damper repeatedly, and play loud music. Wet a rag with raccoon eviction fluid or ammonia and wedge it above the damper. Usually the female will move the kits out within a few days of this harassment. Smoking raccoons and young kits is not recommended, as the young kits are likely to suffocate. Do not cap the chimney until the kits are removed.

It is illegal to release trapped raccoons in many areas, due to the threat of disease. Call local authorities or professionals for removal problems.

DAMAGE ID: Northern Raccoon

PREY ON

Poultry, eggs, rabbits, small lambs

TIME OF DAY

Night

METHOD OF KILL

- Several birds often killed in one night, heads bitten off and carried some distance from the bodies

- Legs, feet, even heads grabbed through fence or chicken wire and torn off

- Crop, breast, entrails chewed or eaten

- Pieces of flesh deposited near water

- Eggs eaten, with shell pieces remaining in nest. Eggs often cracked on sides; nests disturbed

- Egg sometimes moved up to 28 feet before being eaten

- Rabbits and small lambs killed by chewing on nose

- Smudge marks of body oil or fur left on rough surfaces or near dens

- Fences climbed and locks and latches opened into poultry or other enclosures

- Same feeding areas returned to night after night

- Existing human or wildlife trails used, or paths made close to buildings and fences

- Holes dug with refuse and dirt piled in one direction

- Portions of one ear of corn eaten on many plants, stalk often broken; small holes dug in melons; fresh sod rolled up in search of grubs

TRACK

Front 1¾–5⅛ inches long, 1½–3¼ inches wide; rear 2⅛–3⅞ inches long and 1½–2⅝ inches wide. Clearly shows 5 toes but heel and nail tips may not be visible unless in soft sand, soil, or mud. Muddy, human-like prints can be seen on structures.

Front

Rear

SCAT

Blunt, 3–5 inches long and ¾ inch in diameter. Often contain fruit seeds. Due to potentially fatal parasite contamination, **do not handle any raccoon feces** without protection.

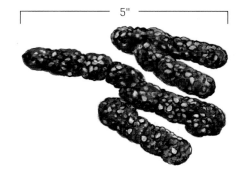
5"

GAIT

Walk is unique and pace-like with same-side front and rear legs moving together; tracks appear differently with various speeds. Stride 8–19 inches. Raccoons also direct-register walk, and bound or gallop when alarmed.

Skunks

Mephitidae

Originally thought to be part of the weasel family, skunks, according to recent genetic information, are a separate family that includes the 4 species of hog-nosed skunks (*Conepatus*), hooded and striped skunks (*Mephitis*), and the 4 species of spotted skunks (*Spilogale*). These species are found only in North and South America. Their cousins, the stink badgers, are the other members of this family and live in Indonesia.

The animal was new to European explorers and colonists, and the word *skunk* came from various Native American names; it is called *zorrillo*, or "small fox," in Spanish. Somewhat resembling the Old World polecats, skunks are also called by that name in some areas; however, the endangered black-footed ferret is the only true polecat in the New World. With the clearing of dense forests, skunks' range expanded, and today they are found throughout North America and south to central South America, in both suburban and rural areas.

Skunks are most famous for their defensive ability to spray an oily, amber-colored, noxious-smelling liquid from their anal scent glands in a mist or stream. Despite this feature, Native Americans and farmers occasionally kept skunks as pets to help control rodents and other pests. In the 19th and early 20th centuries, skunks were also raised commercially for their fur, which was often purposefully mislabeled as sable.

Where legal, the more social striped skunk, with scent glands removed, is still bred and sold as a pet. There is no US-approved rabies vaccine for skunks, so if a pet skunk bites someone it is often euthanatized for testing.

DESCRIPTION

Skunks generally have broad, robust bodies; short, stocky legs with powerful feet and claws for digging; a long snout; and thick, brushy tails.

HABITAT AND BEHAVIOR

Skunks are not territorial; individuals often occupy the same range as other skunks. The striped skunk is very adaptable to differing habitats, including woodlands, rocky or dry areas, brushy grasslands, fencerows near open fields, and even suburban neighborhoods — as long as there is a water source. It may dig its own den but often occupies burrows or tunnels made by other animals, as well as hollow logs, rock and wood piles, culverts, and areas in and under abandoned and neglected buildings. Skunks do not generally roam more than 2 miles from their den.

Spotted skunks prefer more isolated woodlands, grasslands, rocky canyons, or farm fields. They may have several dens and share them with other skunks.

Hunting and foraging. Skunks are omnivorous, and their diet varies according to seasonal availability. They primarily eat insects, grubs and larvae, worms, rodents and other small animals, reptiles, amphibians, birds, and eggs. They also consume plants, roots, berries, nuts, and fungi and occasionally scavenge carcasses. Primarily nocturnal, skunks are often more active at dawn and dusk.

LIFE CYCLE

Except during breeding season, skunks are most often solitary animals. Breeding occurs in early spring, with males often mating with several females as young as 1 year old. Two to 6 or more kits are born in early May, usually remaining with their mothers until about 1 year of age, although they are capable of using their scent glands at 1 week old. Although their life span can range from 7 to 10 years, most skunks do not survive their first year, falling prey to road accidents, trapping and hunting, great horned owls and other predatory birds, and domestic dogs. Most animals avoid confrontations with skunks if possible, although larger predators will kill skunks.

Often returning to the same dens, males usually remain alone. A group of females and a single male may use common dens during winter, where they are dormant sleepers during the coldest weeks, rather than true hibernators. They emerge to feed on warmer days. In warmer climates, skunks remain active year-round.

Skunks have excellent senses of hearing and smell, but very poor vision, allowing them to see only up to 10 feet.

HUMAN INTERACTION

Skunks are highly beneficial in the ecosystem for killing rodents, rabbits, rattlesnakes (skunks are immune to their venom), poisonous spiders and scorpions, and many other insect pests. Occasional encounters or visits by a skunk are not a problem in most cases.

Skunks may eat eggs and prey on poultry hatchings and adults, ground-nesting waterfowl, and domestic rabbits. They will also eat honeybees by scratching at the hive and eating the bees that emerge. They are attracted to garbage, bird feeders, and pet food. They can move into buildings, crawl under porches, and fall in window wells, swimming pools, or garbage cans and be unable to climb out.

Although rabid skunks have been discovered throughout the United States and Canada, most cases occur in the Atlantic coastal states, Kentucky, Tennessee, the upper midwestern states south through Texas, Arizona, and California. In Canada, most cases of skunk rabies are found in Manitoba and Saskatchewan.

LEGALITIES

Striped skunks are not protected in most states or provinces, while spotted skunks are fully protected in some areas. In many areas, it is unlawful to release a trapped skunk due to the threat of rabies. Call local authorities or professionals for assistance and regulations about trapping and relocating skunks.

Keeping a skunk as a pet is illegal in most states and provinces. Where legal, skunks must usually be purchased from a licensed breeder and state or local permits may be required.

Striped Skunk
(*Mephitis mephitis*)

The striped skunk is the most common skunk in North America, appearing in 13 subspecies. Weighing from 4 to 10 pounds and ranging from 21 to 32 inches, including the tail, males are slightly larger than females, and northern skunks occupy the larger end of the size range. While striped skunks have long, strong claws and are excellent diggers, they are very poor climbers. They usually have a black base coat with a white stripe on the head, separating into two stripes along the body, and a mostly white flat, bushy tail. Stripes on the front legs and a white chest patch can also be present. In the wild, mutated forms have a brown or cream base color, and breeders have developed several other color varieties.

▲ Striped skunks have adapted to life in cities and suburbs and expanded their ranges across the Canadian prairie.

▲ Striped skunks are found throughout most of the United States, except the very arid Southwest; and across the southern Canadian provinces, except coastal British Columbia.

Approx. 8 inches at shoulder

Tail 9–14 inches   12–18 inches from tip of nose to rump

▲ Striped Skunk

The spotted skunk is smaller and more secretive than its cousin the striped skunk.

SPOTTED SKUNKS

Western Spotted Skunk
(Spilogale gracilis)

Eastern Spotted Skunk
(Spilogale putorius)

The Western spotted skunk and the Eastern spotted skunk are the smallest members of this family. The Western weighs only 1 to 2 pounds and ranges in length from 12 to 18 inches, including the tail. Eastern spotted skunks are longer, at 13 to 28 inches, and weigh 2 to 4 pounds. Spotted skunks are black or grayish black with a variable pattern of stripes and spots, usually including a large spot on the head, 4 to 6 stripes on the neck and body, and a white-tipped tail. Faster, more agile, and better tree or fence climbers than their striped relatives, spotted skunks are able to enter barn lofts or house attics. They are also more secretive, less populous, and declining in eastern areas.

▶ Western spotted skunks are primarily found in states west of the Continental Divide, but also across Wyoming, Colorado, New Mexico, and western Texas; and southwestern British Columbia.

Eastern spotted skunks are found in the central and interior southern states, and Florida.

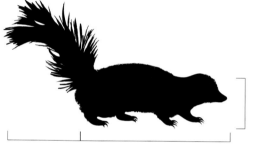

Approx. 5 inches at shoulder

Tail 5–12 inches   7–16 inches from tip of nose to rump

▲ Spotted Skunk

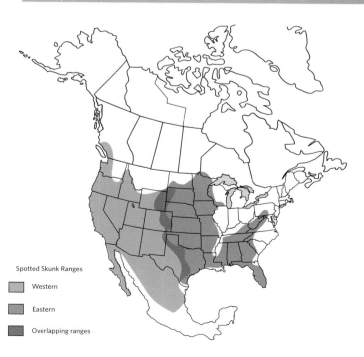

Spotted Skunk Ranges

■ Western

■ Eastern

■ Overlapping ranges

A Most Powerful Weapon

Skunks have a uniquely effective defense system that commands the respect of most other animals.

- Skunks can spray in any position and with accuracy up to 15 feet, generally aiming at the eyes of the predator.

- Skunk spray is nontoxic, although it can be painful and cause burning or temporary blindness in eyes.

- Skunks' loud markings warn other animals to stay away, and therefore they themselves are generally docile, confident, and nonaggressive.

- Before turning to spray a threat, skunks generally give warning hisses, stamp their feet, raise their tail, or make short charges.

- Spotted skunks maintain eye contact while spraying, either standing on their front feet and raising their hind end or arching into a horseshoe shape.

- Striped skunks can discharge their spray 5 to 8 times and then require a week to recharge their glands.

- It is a myth that skunks cannot spray when held by their tail.

- Humans can detect a skunk discharge up to 1.5 miles away.

Skunk Encounters

Due to their poor eyesight, skunks may not see you until you are quite close. Whether the skunk appears alarmed or not, move away quietly and slowly. Do not yell or threaten the skunk since either action will encourage it to spray. Overly aggressive, friendly, or odd movements or an ill or unkempt appearance likely indicate disease and are a cause for great caution. Skunks rarely bite unless cornered or rabid.

ENTRAPPED SKUNKS

If a skunk is trapped in a building, leave all doors and exits open. With gentle encouragement, it will probably leave, although a mother with kits will be most resistant. She will also leave her kits behind while she hunts.

If you suspect that a skunk is occupying a den or other location, first loosely cover the entrance with crumpled newspaper, sprinkle flour or talcum powder on the ground to reveal any tracks, and observe the site for 2 or 3 days. If it remains undisturbed, the space can be permanently filled with gravel or blocked with mesh or other materials. If you are uncertain, you can construct a one-way hinged door to allow the skunk out but not in.

Lights and noise near their denning sites can temporarily repel skunks. Used kitty litter, pepper sprays and solutions, skunk repellents, ammonia-soaked rags, and mothballs can also discourage skunks.

A trapped skunk may be able to climb a rough-covered board or a ramp out of a hole, set at no more than a 45-degree angle. You may also be able to entice a skunk into a cage or box with a smelly food. Cover a trapped skunk since it usually will not spray if it can't see its target.

Dealing with Skunks
Homes and Yards

- Do not feed skunks, since they will lose their wariness around humans.

- Secure all garbage and pet food. Clean up spilled feeds, seeds, fruits, and nuts.

- Feed pets indoors or in a secure pen. Do not leave food out at night.

- Do not turn pets loose when you know skunks are present. Dogs are usually sprayed at night.

- Practice rodent control.

- Water lawns early in the morning so that they are drier by evening. Skunks are able to dig for insects more easily in wet soils.

- Remove brush, rock piles, rubbish, and lumber from areas next to fences and buildings. Fill in potential denning sites, and securely cover access to porches and foundations. Install covers or wire mesh over window wells and other pits.

- Set up variable electronic light and sound guards to deter skunks from areas you use at night. They will habituate to lights left on all night.

Livestock Husbandry

- Secure poultry, waterfowl, and rabbits at night in a predator-proof coop or pen.

- Use livestock guardian dogs (LGDs) to sound an alarm if there is an incursion. They will dispatch a skunk threatening poultry.

- Place beehives 3 feet above the ground to prevent skunk predation. Wrap metal flashing around the base if climbing is a potential problem.

Fencing

- Place wire barriers around footings, buildings, and pens. L-shaped barriers should extend 1 foot downward and outward. Straight wire barriers need to be 2 feet deep. Chicken wire laid on the ground and secured with stakes or pavers will prevent digging.

- Install 2-foot-high hardware mesh or chicken wire around pens and buildings to deter striped skunks from climbing, as well as an electric wire 5 inches off the ground. Deter spotted skunks with 2-foot tall "floppy fences" angled outward and combined with a mesh apron.

Removing Skunk Odor

From pets and other animals

Bathe with foaming mixture of:

- ¼ cup baking soda
- 1 teaspoon liquid dishwashing detergent
- 1 quart hydrogen peroxide 3%

Do not store this solution. Wear rubber gloves and rinse pet thoroughly. Keep out of eyes. Commercial products are also available. Both commercial products and homemade solutions may affect coat color or irritate skin. Tomato juice is not effective at removing odor, only masking the smell.

From surfaces, buildings, or fabrics

- Turn off heating, air conditioning, or other air circulation first. Open all windows and doors, inside and out. Boil white vinegar for 2 hours or so.

- Clean indoor and outdoor surfaces, undersides of vehicles, and laundry with full-strength white vinegar or commercial products.

- Wash color-safe fabrics with oxygen bleach, diluted vinegar or ammonia, or a commercial product specific to skunk odor removal. All are somewhat effective, although it is very difficult to remove the odor completely.

- Place a large pan of boiling white vinegar in a tightly closed vehicle or outbuilding until it cools, then air out the space.

DAMAGE ID: **Skunk**

PREY ON

Poultry hatchings and adults, ground-nesting waterfowl, domestic rabbits, honey bees, eggs

TIME OF DAY

Night

METHOD OF KILL

Though damage occurs at night, striped skunks are occasionally seen during the day. Orphaned and ill skunks are more likely seen during the day.

- Chicks, 1 or 2 birds, or eggs lost
- Animal missing, dragged to den
- Birds and rabbits held against the ground by all 4 paws and killed by a bite to the neck
- Carcass apparently mauled, abdomen eaten; rabbit carcass chewed, but often only the head and foreparts eaten.
- Empty shells or eggs carried up to 3 feet away
- Eggs opened at end, sometimes showing 4 fang marks; shell crushed inward where skunk pushed its nose in; looks as though the chick hatched
- Fertile eggs, closer to hatching, chewed into smaller pieces
- Spotted skunks will throw eggs between their legs against objects to break them.
- Nests torn apart
- Skunk smell present, although it can be carried and left by other sprayed animals (dogs or livestock)
- Surface digging in grassy areas, with small cone-shaped holes or fresh sod rolled back
- Access usually gained through holes, gaps, or by digging
- Lower ears of garden corn eaten

TRACKS

Striped. Front 1⅝–2 inches long, 1–1¼ inches wide; rear 1¼–2 inches long, 1–2¼ inches wide.

 Front

 Rear

Spotted. Front 1–1⅝ inches long, ¾–1 inch wide; rear ⅞–1⅜ inches long, ¾–1 inch wide. Claw marks present, fifth toe and heels might not be visible.

 Front

 Rear

GAIT

Striped: Walk overstep or direct-register, stride 4–9 inches; also trot stride 8–11 inches; lope stride 3–7 inches, often move sideways.

Spotted: Walk overstep or direct-register, stride 2–6 inches; also bound, stride 6–30 inches; or lope, stride 9–11 inches.

SCAT

Blunt, 5 inches long, and ¾ inch in diameter; contains insect pieces.

Striped

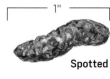
Spotted

CHAPTER 9

Opossums
Didelphidae

Technically, a *possum* is the Australian species and the American species is called *opossum*. From its original home in the Southeast, the opossum has rapidly increased its range north and west, and into cities and suburbs alike.

VIRGINIA OPOSSUM
(*Didelphis virginiana*)

North America's only marsupial, the Virginia opossum originally migrated northward from South America, where some 100 species of opossums and shrew opossums are still found. Early settlers recorded the Algonquian name as *opassom* or *aposoum,* meaning "white dog." *Possum* became a common name in many areas of the country and was also adopted for some Australian marsupials.

The little opossum is well known for its involuntary reaction to a threat known as "playing possum."

Throughout the 20th century, the opossum extended its original range outward from the southeastern United States, moving into southern Ontario by the 1960s and now found in the eastern and central United States. Beginning in 1890, the opossum was introduced into the western coastal states and lower British Columbia — escapees from fur and meat farms, former pets, and deliberate releases. Unfortunately, more recent introductions have occurred in Arizona, New Mexico, Colorado, and Idaho. Highly adaptable to coexisting with humans, the opossum is no longer widely hunted for food or fur and its population has grown.

From his original home in the Southeast, the opossum has rapidly increased his range north and west, and into cities and suburbs alike.

DESCRIPTION

Opossums are small to medium sized; their white, gray, or blackish coats have sparse outer hair but a dense undercoat. The rounded ears and prehensile tail are nearly hairless. The long snout is filled with 50 teeth, more than any other North American mammal. The rear feet have opposable thumb-like digits. Opossums are 14 to 36 inches long and weigh 4 to 14 pounds.

Although they are good climbers and swimmers, opossums forage primarily by ambling along the ground about a half-mile every night. They can use their tail to grasp a tree limb for a support or to carry nesting materials. Contrary to many depictions, the adults do not hang by their tails, although very young opossums may do this briefly.

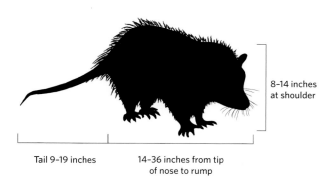

Tail 9–19 inches   14–36 inches from tip of nose to rump

8–14 inches at shoulder

▲ Virginia Opossum

HABITAT AND BEHAVIOR

Although opossums prefer deciduous woodland near water, they are also found in agricultural fields, grasslands, swamps, and more arid areas and are very adaptable to suburban and urban areas as well. Not well prepared for cold weather below 20°F for extended periods of time, opossums often suffer frostbite to their tails and ears. Harsher winters limit their presence into northern and mountainous areas.

Opossums may spend daylight hours in trees using cavities and crevices and bird or squirrel nests. They also use dens made by other animals, especially during colder months, and they carry in material to make a messy nest. They can also be found under porches; amid debris or brushy piles; or in crawl spaces, buildings, and drains.

Hunting and foraging. Nomadic omnivores, opossums move randomly through an area during the night, searching for food. Invertebrates such as insects form a large portion of their diet, but they also consume grasses, fruits, seeds, nuts, grain, and small animals such as mice, birds, amphibians, reptiles, and eggs, and they scavenge carcasses, garbage, and animal feed as well. Since they do not hibernate they must search for food whenever possible during the winter, even during daylight hours.

Solitary and nonterritorial, females may stay in one area as long as there is sufficient food, while

▲ Opossums range from the eastern United States to the mountainous West and are also found in Southern Ontario and the West Coast into British Columbia, with isolated introductions elsewhere.

young animals and adult males tend to roam freely. Opossums will use human and wildlife trails, roads, or move along buildings.

LIFE CYCLE

Females are capable of breeding almost year-round, depending on climate. They have 2 or rarely 3 litters, each averaging 4 to 8 surviving young, or **joeys**. Opossums are marsupials; the embryos crawl into their mother's pouch, where they nurse for about 60 days before emerging.

The young opossums are independent at 2 to 3 months of age and sexually mature a few months later. They grow continuously throughout their short lives. Most survive only 1 to 2 years; opossums only rarely reach the age of 7. Coyotes, foxes, bobcats, dogs, raccoons, raptors, and owls will prey on opossums. Many die through road accidents while scavenging roadkill. They are partially or completely immune to the venom of rattlesnakes and other pit vipers.

HUMAN INTERACTION

Opossums are valuable predators of insects, snails, slugs, maggots, and rodents, and they consume carrion. Although they are nonaggressive, they may nest in garages, attics, and outbuildings; under porches; and in brush or rubbish piles — defecating or damaging areas in buildings. They may also eat pet food and garden vegetables and will disturb unsecured garbage.

Although they are not major carriers of rabies and the disease is not present in some entire state populations, opossums do carry the parasite that causes fatal equine protozoal myeloencephalitis (EPM) in horses. Horse owners should exercise prevention methods (see page 164) and carefully clean any feed containers or surfaces contaminated by scat, since the parasite remains viable up to 1 year.

LEGALITIES

Different states have specific hunting or trapping seasons. Consult wildlife authorities before relocation, which may be prohibited due to threat of disease.

Opossum Encounters

Opossums may hiss, growl, or show their teeth at a threat but they usually use a different tactic: they "play possum" (see box, page 164).

Opossums may become aggressive if cornered, especially by a dog. If you find orphaned or young opossums, call local wildlife rehabilitators, as it is very difficult to keep this species healthy in captivity.

Entrapped Opossums

Opossums are easier to evict than many other animals because they tend to move after 2 or 3 nights unless the site is very attractive or babies are in the pouch.

- Use flashing lights and loud music to encourage them to move on.

- Set out rags soaked with ammonia, or place ammonia in a can with a punctured top.

- Leave building doors open at night to allow opossums to leave.

- Bait a live trap with smelly food. Release a trapped opossum outside, but do not transport it.

- Use one-way doors and tracking powder to make sure the animal has left and then seal the space.

- If an opossum has fallen into a trash container, tip the container over and let the animal walk away.

Dealing with Opossums
Homes and Yards

- Do not feed wild opossums, because it will encourage familiarity and threatening behaviors and will attract more opossums to your yard.

- Remove attractants such as birdseed and fallen fruit. Secure garbage and compost containers. Bury food scraps in compost piles.

- Do not leave pet food out at night. Lock pet doors.

- Clean outdoor cooking areas thoroughly.

- Clean up brush and other refuse, and trim tree branches over roofs, animal housing, and pens.

- Fill old animal burrows and tunnels with gravel or cover with mesh.

- Close vents and gaps, cap chimneys safely, and cover access to areas under porches and crawl-spaces in buildings.

- Eighteen inches of sheet metal and metal or plastic spikes will prevent climbing up buildings, trees, and poles.

- Note that floodlights are not always effective, although opossums may avoid well-lit areas. Electronic motion lights and sprinklers may be more useful around the home.

Livestock Husbandry

Secure poultry and rabbit housing and pens by covering all openings with ¼-inch hardware cloth. Livestock guardian dogs (LGDs) will also protect against opossums.

Fencing

- Reinforce existing nonmesh fencing for yards, gardens, orchards, and animals with 2 low electric scare wires at the bottom and another at the top.

- Install exclusion fencing for animal housing, pens, waterfowl, and game birds. Use 3- to 4-foot woven wire with 3-inch spaces.

- Prevent climbing by installing a top scare wire 3 inches out from the fence.

- Bury an exclusion barrier 4 to 6 inches down and 8 to 12 inches outward.

Playing 'Possum

When threatened, opossums often fall into a seemingly unconscious or catatonic state in which they appear dead or very sick — even though their metabolic processes continue unchanged and they are potentially capable of action. The unresponsive animal may lie on its side, its mouth open and its teeth exposed, with foaming drool or saliva, for a few minutes or up to 4 hours. The eyes may be partially or completely closed, and foul-smelling fluid is expelled from the anal glands. When ready to rouse, opossums usually move their ears first or raise their head to gauge any threats. Very young opossums often cannot play 'possum.

Never touch an unresponsive opossum without heavy gloves, and use great caution if attempting to move it. Opossums may become aggressive if cornered, especially by a dog.

DAMAGE ID: Virginia Opossum

PREY ON

Poultry, waterfowl, game birds, young rabbits; will eat eggs; scavenges carcasses killed by other predators

TIME OF DAY

Night

METHOD OF KILL

▸ A single sleeping bird or chick killed

▸ Bites left on the body or an entire bird consumed on-site, leaving only a few wet feathers

▸ Head eaten first or vent chewed through to eat internal organs, preferred over muscle meat

▸ Eggs messily mashed, with small shell pieces left in nest

▸ Opossums often linger in the vicinity after eating to sleep and will return nightly to a food source.

TRACKS

Front 1–2¼ inches long, 1¼–2½ inches wide; rear 1¼–2¾ inches long, 1½–3 inches wide. Five long toes on the rear footprint may resemble fingers with an opposable thumb, pointing into the track. Claws may register. Can leave muddy prints where they climb and a dragging tail print.

Front

Rear

SCAT

Variable in shape and size, often pointed ends, 1–4 inches long and ½ inch in diameter

OTHER SIGN

Smooth wear marks or long silver or gray hairs at entrances — burrows, entry holes in buildings, or under a fence

GAIT

Walking stride 4.5–9 inches, pacing or moving front and rear legs on same side simultaneously, may overlap tracks; trot stride 10–15 inches

CHAPTER 10

Rats

Muridae

The Muridae family of Old World rats are now found everywhere around the world except Antarctica. These rats and mice can be both serious pests and useful laboratory animals. Genus *Rattus* ("true rats") includes the brown or Norway rat, black or roof rat, and the Polynesian rat. Several species of the genus *Sigmodon*, the cotton rats, are native to North and South America.

The Old World rats were introduced to North America during the earliest days of exploration and colonization. Although there are native rats such as the wood or pack rat, cotton rats, and kangaroo rats, the invasive rats are by far the most common and troublesome rodents to homeowners and farmers.

HABITAT AND BEHAVIOR

Norway and black rats live around residential, industrial, and farm sites; cropland; and parklands, especially near ponds, rivers, and waterways. Rapid breeders, they quickly infest an area. They often transmit bacteria, viruses, parasites, and other diseases that can contaminate food and buildings.

To deal with their constantly growing teeth, rats must continually gnaw. They use their oversized front teeth to chew foodstuffs and housing or bedding materials, often causing great damage to structures.

Although they have poor eyesight and are colorblind, rats use their good senses of hearing, smell, taste, and touch to locate food. Capable of breeding year-round in warmer areas, they produce 5 to 8 pups per litter and 4 to 5 litters per year.

Hunting and foraging. Rats are usually nocturnal, beginning to search for food after sunset. They will hoard food to eat later, caching it in a variety of locations. Rats are omnivores, eating fruits, seeds, insects, snails, garbage, foodstuffs, and animal feed. As opportunistic animals, they will also eat eggs, meat, and fish. They require daily access to water.

HUMAN INTERACTION

Rats contaminate food; spread disease (plague, salmonella, leptospirosis, and tularemia); inflict serious damage to buildings and wiring; prey on poultry, eggs, and newborn animals; and destroy crop and landscape plants.

LEGALITIES

Norway and black rats are not protected, and any legal means may be used to control them.

Dealing with Rats

While cats cannot usually hunt rats successfully, the ratting dog breeds are dedicated to the pursuit of rats and have long been welcome in barns and farmyards. No electronic frightening or ultrasonic devices have long-term effectiveness, as rats become habituated to them.

Any lethal control of rats will not solve an established problem unless you implement prevention methods to control what is attracting them. Any use of poisons carries a danger to children, pets, livestock animals, and other species. All rodent baits should be considered toxic to dogs, cats, and wildlife. Rats are leery of new objects and will often avoid traps or bait stations for several days.

- Store animal feed in containers. Eliminate free access to animal food in paddocks and pens if possible.

- Feed dogs and cats inside or remove food after they eat.

- Prevent access to bird feeders, and clean areas under them.

- Dispose of all garbage and trash, including fallen fruit and garden produce.

- Use covered compost containers.

- Remove unprotected water sources wherever possible, as water is a strong attractant to rats in hot or dry weather.

- Clear vegetation and debris near buildings. Stack all wood and building materials 1 foot away from walls and 18 inches off the ground.

- Use plastic or metal baffles and barriers to prevent rats from climbing trees, pipes, and so on. Remove overhanging tree branches and vines, which give rats access to animal pens and structures.

- Close all doors tightly, including garage doors.

- Rat-proof buildings and pens. Use ¼-inch hardware cloth to cover all gaps. Use solid concrete flooring in animal pens or feed storage areas with concrete rat walls, treated wood, or hardware-cloth aprons to prevent digging into buildings and pens.

Norway rats are highly adapted to living closely in and around human buildings in both city and countryside.

Norway Rat
(*Rattus norvegicus*)

The Norway or brown rat, also called the sewer or wharf rat, is not a native of Norway, despite the belief that it arrived in England from that country. Sturdy and comparatively large, Norway rats weigh up to 18 ounces. They can be as long as 16 inches, including the nearly hairless tail, which is slightly shorter than the length of the head and body. They are primarily grayish brown although black, white, and mottled rats are seen, and they have small ears and a blunt snout.

Norway rats are adaptable to most climates. Usually found on ground or lower floors of buildings, they live primarily in burrows.

▲ The Norway rat is found in the entire continental United States and Canada except Alberta. Alberta has successfully and proactively eliminated rats for more than 50 years.

Black Rat
(*Rattus rattus*)

The black or roof rat is similar in length, but its hairless tail is longer than the Norway's, and its body is slenderer and darker. Black rats weigh up to 10 ounces. Their ears are large, and the face has a pointed snout. Excellent climbers, they are more likely to be found in trees and rafters and on wires and roofs than are Norway rats.

Black rats are limited to warmer climates around coastal cities, where they frequently enter buildings through the roof and live in utility spaces and false ceilings or attics, as well as in trees, brush, and ivy.

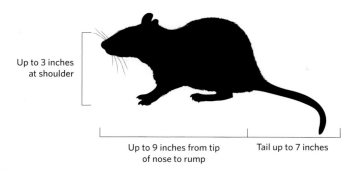

Up to 3 inches at shoulder

Up to 9 inches from tip of nose to rump

Tail up to 7 inches

▲ Norway Rat

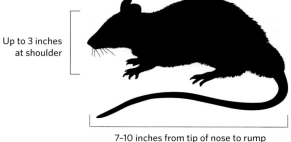

Up to 3 inches at shoulder

7–10 inches from tip of nose to rump
Tail longer than body

▲ Black Rat

▲ Black rats can be distinguished from their Norway cousins by their larger ears and pointed snout.

▲ The black rat is limited to the lower half of the eastern coast, the Gulf states into Texas, and the entire Pacific coast. Isolated groups may be found elsewhere at times.

Cotton Rat
(*Sigmodon hispidus*)

These rats were named for their habit of building nests out of cotton. Native to North and South America, cotton rats range from southern Virginia through the Gulf Coast states; west to northern Missouri, Nebraska, and Kansas, through southeastern Colorado, New Mexico, and southeastern Arizona and California.

The cotton rat is 10 inches long, including its bare tail. It is smaller than the Norway rat, with a shorter tail, grizzled black or gray fur, and large ears hidden in fur. Active day and night, cotton rats are herbivores that eat green vegetation, quail eggs, and more rarely insects or small animals. They will damage grass, alfalfa and grain crops, fruits, berries, nuts, and vegetables such as peanuts, sweet potatoes, and melons. Cotton rats cut plants off at the base, chop them into smaller pieces, and leave small pieces of grass or stems in piles along their grassy runs. Nests are found in grassy or overgrown fields, ditches, or fencerows. They will also infest outbuildings.

Solid sheet-metal barriers 18 inches tall and buried 6 inches deep will exclude cotton rats. They are a host for airborne hantavirus; handle droppings carefully and wear a protective mask.

DAMAGE ID: **Rats**

PREY ON

Poultry, eggs, newborn animals

METHOD OF KILL

- ▸ Chicks or eggs missing, carried off at night
- ▸ Chicks or larger birds stuck in a rat tunnel
- ▸ Birds chewed up, legs often bitten
- ▸ Newborn pigs, lambs, or calves bitten
- ▸ Smell of urine and scat; musky rat odor
- ▸ Gnaw marks on wooden surfaces
- ▸ Sounds of gnawing, scratching, fighting, and squeaking
- ▸ Fresh black scat, turning gray over time
- ▸ Narrow pathways or trails found along walls or fences
- ▸ Dirty, oily rub or smudge marks left on walls or floors along paths
- ▸ One or more smooth, hard-packed tunnel entrances to the burrow, 2–4 inches in diameter, with fresh soil at opening; concrete slabs burrowed under

TIME OF DAY

Night

TRACK

Front ½–⅞ inch long, ½– ¾ inch wide; rear ¾–1⅜ inches long, ⅝–1¼ inches wide.

GAIT

Walking stride 3–5 inches; bound stride 6–21 inches

SCAT

Small unconnected ovals, ¼–1 inch long, ⅛–¼ inch in diameter

Roof Rat

3/4"

Norway Rat

CHAPTER 11

Domestic and Feral Animals

Although we assume wild animals are the major predators, in fact domestic animals are responsible for significant losses of stock and poultry. In the case of swine, these animals are truly feral or out of human control. Unfortunately, domestic pet dogs, not feral dogs, are guilty of killing or injuring very large numbers of stock and poultry.

The public is often surprised to learn that roaming dogs are the second most common killer of sheep and cattle in the United States. Roaming pet dogs also chase and injure stock, including equines, and kill poultry. Truly feral dogs threaten pets as well, and street dogs in more urban areas are responsible for attacks on people.

Roaming and feral cats can seriously damage wild bird populations, despite their beneficial rodent hunting. In many areas, roaming or feral cats are a great threat to endangered species.

In addition to attacking stock, both feral domestic pigs and wild boars can cause severe environmental and crop damage.

SWINE
(*Sus scrofa*)

In North America, most so-called wild hogs are **feral** pigs, the descendants of domestic pigs, released or escaped into the wild since the earliest days of exploration and colonization in the 16th century. Found throughout much of the countryside, they were called razorbacks, pineywoods, and woods hogs. Lacking predators in many situations, they rapidly reproduced and populations exploded.

Beginning in the early 20th century, true wild boars from Europe were intentionally released in many states for hunting. In the 1980s and '90s, in both the United States and Canada, additional European boars were released or escaped from hunting preserves. These wild boars and their hybridized crosses with domestic pigs are found in some areas. The population of feral pigs is estimated at 4 to 5 million in the United States and 1 to 3 million in Canada.

2½–3 feet at shoulder

5–6 feet from tip of nose to tip of tail

▲ Feral Pig

DESCRIPTION

Despite feral pigs' various breed ancestries, survival tends to favor a leaner, wilder form rather than the appearance or bulk of many domestic pig breeds. Most feral pigs are rangy in appearance, with long legs, long snouts, and darker colors. Colors vary, depending on the amount of true wild boar in their genetics, including the most common black, followed by spotted black, reddish brown, and black and white. Also seen are brown, black and brown, white, and roan. Young pigs are usually reddish brown with horizontal stripes, eventually darkening from brown to black.

Feral pigs are often thick-skinned, with a dense outer coat of coarse hair or bristles, a wooly undercoat, and a 5- to 6-inch mane of hair on their neck and shoulders. Usually seen walking or trotting, they can run up to 20 mph. Armed with continuously growing 8- to 9-inch tusks, feral pigs generally weigh up to 250 pounds, although individuals can be much larger. Pigs have excellent senses of smell and hearing, and almost panoramic vision.

HABITAT AND BEHAVIOR

Feral pigs are highly adaptable to habitats from tidal marshes to hardwood and conifer forests to mountainous areas below the snowline where temperatures remain above freezing. True wild boars can survive in more mountainous and colder areas. Although feral pigs prefer marshy, brushy, and forested areas, they will feed in open areas and pastures,

▼ Feral pigs can be found in many colors, color patterns, or sizes — although black and reddish brown are common.

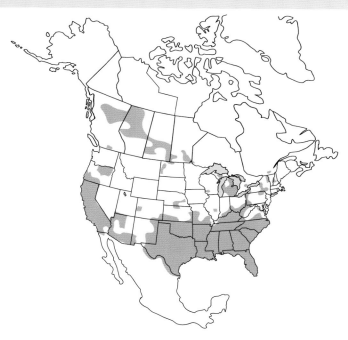

▲ Feral pigs are found across the south-southeastern United States, north into the lower Midwest, west to California, and in Hawaii. Texas and Florida have the largest populations, about 1 million animals each. Smaller pockets exist in northern states outside the typical range. Feral hogs are also found in Ontario, Manitoba, Saskatchewan, Alberta, and British Columbia.

generally at night, and will seek ponds and streams for cooling during hot weather. Typically active during the day and evening, feral pigs often shift to nocturnal feeding during hot weather or in response to hunting pressure. They sleep together in den areas, typically found in shade or brush.

As omnivores, pigs have a varied diet. Favored vegetation includes grasses, plants, roots, tubers, fruit, and seeds. Acorns and hickory nuts are especially attractive. Pigs also eat earthworms, insects, amphibians, fish, crabs, snakes, turtles, rodents, eggs and chicks of ground-nesting birds, fawns, and young livestock.

LIFE CYCLE

A few related sows and piglets often live together in a group called a **sounder**, remaining in a large home range of variable size depending on habitat and food, moving around within the range seasonally to follow food sources. Individual sows with piglets or

other individuals, mainly adult males, are also commonly seen. Larger groups sometimes form where there is a good food source.

Several boars will seek out a female in estrus, and fighting will occur. Sows can reproduce year-round, beginning at 6 months of age and producing up to 2 litters yearly. They can live 20 to 30 years. Predators of young pigs include mountain lions, bears, alligators, and wolves, although human hunting is responsible for a significant mortality of feral hogs.

HUMAN INTERACTION

Rooting in the soil for food, causing erosion, and preying on native species, feral hogs cause enormous environmental and agricultural damage. In addition to crop damage, feral pigs prey on lambs, kids, and newborn calves, biting or crushing the head or neck. Attracted by smells on birthing grounds, they often target weaker or smaller animals such as twin lambs. They will also chase flocks, which may appear agitated.

It can be difficult to determine whether swine caused a death, because they usually consume the entire carcass, also feeding on stillborn animals and afterbirth. The ground near an attack is often disturbed and trampled, making it hard to locate tracks and other signs. Deaths are often attributed to coyotes where their ranges overlap.

Feral swine carry 45 viral or bacterial diseases and parasites, all of which can be transferred to humans, livestock, and pets through contact with bodily fluids, handling, and ingestion of tissues. Take great care when handling dead pigs or animal carcasses killed by pigs. Do not feed remains to dogs. Legal disposal varies by state.

LEGAL ISSUES

Feral hogs are considered game animals in some states. Contact your state wildlife agency or agriculture department to determine need for a permit. In most other states, feral hogs are unprotected and the landowner may authorize hunting or trapping. The US Fish and Wildlife Service hunts and traps feral hogs for control. Regulations vary by province in Canada.

Dealing with Feral Swine

- Frightening methods are not effective.

- Use livestock guard dogs to discourage and harass feral pigs.

- Install fencing around garden and orchard areas to keep feral pigs out of your property.

- Set up exterior electric scare wires to reinforce existing fencing and prevent digging.

- Construct perimeter fencing from high-tensile electric wire or woven wire, electric hog wire mesh, or hog panels with a height of at least 40 inches.

- Space electrified wires at 6, 10, and 18 inches above the ground to deter pigs. To exclude baby pigs, space wires 1.5 to 2 inches apart near the bottom.

DAMAGE ID: Feral Swine

PREY ON

Lambs, kids, newborn calves

METHOD OF KILL

▸ Entire carcass consumed or carried off; tracks and blood evident, along with general trampling of area

▸ If not fully eaten, carcass skinned off and rumen and stomach contents consumed

▸ Skull or neck bitten or crushed in the attack

If feral pigs are present in your area, you will see signs:

▸ Rooting damage to forest, bottomland, pasture, or cropland

▸ Muddy wallows near water

▸ Muzzle impressions

▸ Dried mud and hair on trees, posts, logs, or large rocks

▸ Well-worn trails with signs of hair or dried mud

▸ Shallow dirt beds or nests in shady or brushy areas

TIME OF DAY

Day and evening; nocturnal in hot weather or near humans

TRACKS

Front 2⅛–2⅝ inches long, 2¼–3 inches wide; rear 1⅞–2½ inches long, 2–2¾ inches wide. Two-toed footprints can resemble those of a deer or calf, but are more blunt or rounded. Dewclaw impressions may be visible. Track varies with degree of wild boar influence.

 Front

 Rear

GAIT

Walk stride 25 inches; trotting 55 inches; lope or gallop stride 36–52 inches

SCAT

Tubular or pellets, depending on diet

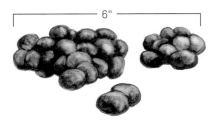
6"

DOGS
(Canis familiaris)

True **wild dogs** are neither domesticated canids nor their descendants. There are no wild dog species in the United States or Canada. Feral, stray, or street dogs and roaming pets are a troublesome problem in both city and country. Roaming dogs are a serious danger to livestock and poultry on farms and ranches.

Worldwide, true pariah dogs are medium in size and coat, often red or yellowish, with erect ears and curly tails.

Several terms are used to describe the different ways domestic dogs behave outside human control:

- **Pariah dogs** is a term used in ecology to describe dogs that breed freely for generations but are dependent on human handouts or refuse.

- **Feral dogs** are born and exist outside human control and are not necessarily dependent on human food sources.

- **Stray dogs** are born into domesticity but became lost, roamed off, or were abandoned. They are almost completely dependent on human food sources.

- **Street dogs** in urban environments can be a combination of stray or feral dogs.

- **Roaming dogs** are pets that have temporarily escaped or are allowed to range freely.

Pariah Dogs

Sometimes called primitive dogs, strains of pariah dogs can be genetically distinct. In the southeastern United States, Carolina dogs may be the descendants of free-ranging primitive dogs, but they are not living like pariah dogs close to human settlements and dependent on refuse or handouts. There are no populations of true pariah dogs in the United States or Canada.

Pariah dogs do not tend to establish extended family groups of related individuals, especially in urban areas. Where food supplies are plentiful, such as around refuse dumps, large groups will congregate in small areas. Pariah dogs in these situations can be quite comfortable around humans, while others may be unapproachable.

Feral Dogs

Locally, feral dogs will retain the basic appearance of the common dogs in the area. They often resemble German shepherd or husky types, although hunting dogs, pit bulls, and other similar types are more typical in some areas. For feral dogs, survival tends to favor animals that are neither too small nor too large.

Truly feral dogs can have a very large range but are far less common in remote areas where large predators are present. Feral dogs are also found on military reservations, large airports, or similar sites. Established populations of feral dogs are not as common in agricultural areas due to the low tolerance by landowners, who tend to eliminate them.

Although they may be solitary, feral dogs in rural areas often live in small, interrelated extended family groups with established denning sites and trails. The size of the group and its range is dependent on the availability of food. They seek den sites in brushy or forested areas or near abandoned human buildings and refuse. Dens may also be dug or appropriated from abandoned wild animal dens. Around the site will be evidence of prey consumption.

Feral dogs in more rural areas are opportunistic and will adopt a diet similar to coyotes', consisting of small animals, ground-nesting birds, and vegetation. They hunt or forage during the night as well as at dusk and dawn. Established feral dogs can kill prey in the same way coyotes and wolves do, and they will consume what they have killed or return with it to their den. Truly feral dogs are fearful, wary, secretive, and potentially aggressive due to lack of socialization.

Stray and Street Dogs

Stray dogs can have any appearance and may appear anywhere; however, they are more common in urban areas where food is more available. Urban areas report large numbers of dogs cast off from dog-fighting activities, including both the fighting or breeding dogs and diverse bait dogs. Large cities in the United States estimate their street dog population as high as 50,000. Stray dogs in rural areas may survive on the fringes of residential areas or farms, but they are generally unwelcome and do not tend to form large groups.

When Good Dogs Go Bad

Roaming domestic dogs chase and injure stock for fun, not for food. Owners are often shocked that their beloved and friendly pet killed farm animals or poultry. In addition to establishing good fencing and using livestock guardian dogs (LGDs), owners can protect their animals by practicing good prevention strategies:

- Do not allow unsupervised contact between your own pet dog and your stock, especially if the dog is from a breed with a high prey or chase drive.

- Do not let your dog play with neighbor or roaming dogs, which may lead to dangerous play behavior with your stock.

- Be very cautious of visiting dogs, and strongly discourage neighbor dogs from entering your property.

- If you use an LGD, discourage anything more than cursory familiarity between your LGD and your pet dogs. Definitely do not allow neighbor dogs to visit your property and interact with your LGD. You want to maintain your LGD's defensive aggression against a roaming dog that might easily threaten your stock.

What to Do When a Domestic Dog Attacks Your Animals

Dog attacks are often very emotionally disturbing because of the violence and widespread destruction or mutilation of your animals. It is helpful to plan ahead and know which authorities to contact in your area, as well as what further actions to take.

If you catch the dog or dogs in the attack, take the following steps immediately:

1. If the dogs are friendly, restrain them and call the animal authorities rather than their owners. Take photos of the dogs before releasing them to the authorities. Record all identifying information from tags.

2. If the dogs run away, try to follow them calmly. They are less likely to run if you don't chase them. Take photos. Most dogs will return home. Do not confront the owners who may deny their dog's involvement, remove evidence through bathing, or create an alibi before the authorities arrive. Call the authorities.

3. If you can't restrain or follow the dogs, write down an accurate description of their sizes, ears, tails, coat color and length, markings, and so on. Again, take photos if possible.

4. Take more photos to document the full extent of the attack.

Call the authorities even if you come upon the attack after the dogs have gone. Even if they cannot establish ownership, you have a record of the attack and may be eligible for monetary compensation from local government sources or insurance.

In some areas, owners of stock have the right to shoot a dog that is attacking their animals; however, the laws differ in various states, provinces, and municipalities. If you shoot someone's pet, you may find yourself involved in a legal action or lawsuit.

Street or stray dogs often live solitary lives or form fluid and changing social groups of 2 or 3 individuals but do not tend to establish extended family groups of related individuals. Where food supplies are plentiful, large groups will congregate in small areas. Females reach sexual maturity as early as 7 or 8 months and are capable of producing litters every 6 to 7 months. Survival of pups is low, and numbers are mostly replenished from newly abandoned or stray dogs. Some urban street dogs may respond favorably to humans, while others will be extremely skittish, frightened, or potentially dangerous.

Outside urban areas, stray dogs scavenge garbage and refuse sites near humans and eat carrion. Hunting behavior is not common and when attempted is unskilled.

Roaming Pet Dogs

Roaming dogs may be of any appearance or breed. Their reaction to humans can be friendly or calm. Roaming dogs are responsible for most damage to livestock and poultry in rural areas.

HUMAN INTERACTION

Dogs are documented as the second most common killer of sheep, lambs, cattle, and calves in the United States, responsible for thousands of deaths each year. Feral, stray, and roaming dogs may all prey on livestock or poultry, but stray and roaming dogs do the most damage, killing or seriously wounding many animals, as they are either unskilled at hunting or are not seeking prey to eat. These attacks are usually more extensive and expensive for the owner.

Feral dogs also kill pet dogs or cats and may attack people, especially children. Urban areas report a very large number of bites by street dogs. Stray or feral dogs can transmit rabies, disease, and parasites.

LEGALITIES

Local and state laws vary. In some areas, farmers or ranchers are allowed to shoot dogs chasing or killing livestock or game animals such as deer. Urban areas have animal control agencies responsible for removing stray or feral dogs.

Dealing with Feral or Stray Dogs

Homes and Yards

- Do not feed stray or feral dogs.

- Eliminate hiding or denning areas.

- Remove animal food, and eliminate garbage and carcasses that will also attract dogs.

- Do not allow neighboring or stray dogs on your property. Do not allow your dogs to play with stray dogs.

- Employ an LGD to discourage stray or neighbor dogs.

- Use pepper or bear spray to deter a personal threat.

Livestock Husbandry

- Scare devices with sound and lights can discourage feral dogs, although domestic dogs are not frightened by artificial lighting, human sounds, or activity; in fact, they are often attracted to well-lit barnyards.

- Livestock guardian animals are reliably aggressive toward dogs, although llamas and donkeys may also be victims of attacks.

Fencing

- Use scare wires on the top and bottom of fences for reinforcement.

- Install exclusion fencing, either wire mesh (no spaces larger than 4 × 6 inches) or electric.

Coyote, Feral Dog, or Roaming Pet?

Coyotes and dogs are the leading killers of stock, and their attacks are easily confused. Dog attacks, however, generally involve indiscriminate and extensive mutilation and deaths.

If you determine that a dog or dogs caused the attack, remember that roaming pets or stray dogs, not truly feral dogs, commit most livestock or poultry damage. A major Australian study revealed that 40 percent of attacks are by a single dog. In 51 percent of cases a pair of dogs committed the attack together — usually a male and female or 2 dogs who live together or next door to each other.

Only 9 percent of the attacks involved 3 or more dogs. One or 2 dogs can do such extensive damage that it often seems that there must have been a pack of dogs involved.

It is vital to determine whether the dog simply bothered stock or ate an already deceased or stillborn animal, rather than killing it. LGDs will sometimes try to eat dead animals or afterbirth to prevent attraction of predators. Check the field autopsy damage ID guides on pages 18–21.

Exceptions to the ID guide below can occur in summer when a coyote is teaching her pups to hunt. The pups may attempt several, inexperienced attacks on stock. Truly feral dog attacks can resemble coyote attacks.

	Sign	Coyote	Feral Dog	Roaming Pet
Time of Day	Dawn		✔	✔
	Day		✔	✔
	Night	✔	✔	✔
Tracks	Open or splayed		✔	✔
	Oval or rectangular	✔		
	Toes pointing in different directions		✔	✔
	Toes are closer together	✔		
	Tracks clearly left by 2 adult animals			✔
	Follow a straight line	✔		
Scat	Brown			✔
	Black	✔		
	Smooth, soft			✔
	Contains pieces of hair and bone	✔		
Length of Attack	Attacks usually last a long time			✔
	Animals killed quickly and efficiently	✔		
Condition of Animals	Surviving animals stressed and traumatized		✔	✔
	Prey chased every time they move			✔
Method of Kill	Many individuals attacked		✔	✔
	Young, weak, or injured individuals attacked	✔		
	Animal attacked from side, with rips or bites in head, neck, shoulders, flank, or hindquarters			✔
	Wounds appear slashed or ripped			✔
	Prey completely carried away	✔		
	Kill eaten	✔		
	Predator returns to feed on carcass	✔		

DAMAGE ID: **Domestic Dog**

PREY ON

Livestock, poultry, pets

TIME OF DAY

Day or night

METHOD OF KILL

▸ Inefficient and prolonged period of attack resulting in exhausted and highly stressed stock

▸ Many animals injured or killed; both young and old attacked

▸ Wounded animals often mutilated with chewed ears, wool, skin, or feathers pulled out and scattered

▸ Attack wounds on side of the animal; rips or bites on the head, neck, shoulders, flank, or hindquarters; wounds appearing slashed or ripped; legs broken or blood on animals' hind legs

▸ Carcass not carried away, covered, or extensively fed upon

▸ Carcass not returned to

TRACKS

Dog tracks are more open or splayed than coyotes', with thick or blunt claw marks, and toes pointing in different directions. Size varies with breed. Coyote tracks are more oval or rectangular; toes are closer together, with less prominent nail impressions or only middle toenails visible.

Front

Rear

GAIT

Stride will vary with size. Gait may vary with breed or body size or conformation. At trot, may overstep, step to side, or direct-register. Less likely to follow straight lines or direct-register than coyotes.

SCAT

Dog scat is usually brown, smooth, soft from kibble foods, and lacking hair or bone. Size varies with breed.

DOMESTIC CATS
(Felis catus)

True wildcats are the ancestors of domestic cats and are native to Europe, Africa, and Asia. There are no wildcat species of genus *Felis* in North America. The domestic cat, although it has lived with humans for about 4,000 years, retains a very strong hunting instinct and can still revert to a free-living life.

Specific terms are used to describe domestic cats when they are outside of human control.

A **stray** cat is born into domesticated life but is either abandoned or lost. A stray cat may seek human interaction and allow itself to be handled.

A **feral** cat may have had no connection to domesticated life for many generations or may be born from a stray mother. Unlike a stray, a truly feral cat will be too fearful to be handled or tamed — although very young feral kittens can be socialized. Estimates of the feral cat population range from 50 to 100 million in the United States and Canada.

A **free-ranging** cat may be domestic, stray, or feral. About 40 million of the country's 77 million domestic pet cats are allowed outside part-time or full-time to free-range.

Feral, stray, and free-ranging domestic cats number in the millions in North America. They can prey on wild and domestic birds, small rodents, and rabbits.

DESCRIPTION

Feral cats do not look significantly different from domestics, although they may be slightly smaller at 4 to 8 pounds and 8 to 12 inches tall. In response to selection pressure, many multigenerational feral cats are shorthaired and found in colors that provide some camouflage; however, since so many cats are strays or free-ranging, coat and color do not provide a reliable identification tool.

HABITAT AND BEHAVIOR

In many suburban or urban areas, feral cats live around human structures and abandoned areas, where they scavenge garbage or refuse and are sometimes fed by people. These semiferal cats may congregate in groups or colonies.

In more rural areas, feral cats usually live entirely independent of humans for shelter or food. Often found near fields, in windbreaks or ditches, they live entirely on what they are able to hunt and are extremely wary of people. Some feral cats will live in and around outbuildings on farms if not chased off by dogs.

Feral cats maintain home ranges whose size depends on the availability of food and denning sites. The territories of male cats are usually larger than females'. Den sites are often hollow logs or tree limbs, old burrows from wild animals, debris, or dense grass. Unlike domestic cats, feral cats do not tend to bury their scat, using it to mark their territories. In addition, they urinate, scent-mark from glands on the face and anal region, and claw logs or trees around their boundaries.

Although they hunt alone, feral cats often live and socialize in a group made up of several adult females that may be related, their offspring through about age 7 months, and an adult male. Young males leave or are driven from the group as they achieve sexual maturity. Feral cats' life span is much shorter than that of domestic cats, especially for kittens. Extremely prolific, females can reproduce at 7 months

of age and can produce up to 3 litters yearly, primarily in spring and summer months. Litters range between 2 and 10 kittens.

Feral cats are most active at night, with the peak of hunting during dusk and before dawn. Cats are also active in the day, especially free-ranging domestic cats. Opportunistic hunters, they consume 5 to 8 percent of their body weight daily from prey such as small rodents, rabbits, birds, reptiles, amphibians, insects, fish, and carrion. In an urban or suburban environment, feral cats may serve as the apex predator, while in other areas they are prey for foxes, coyotes, wolves, mountain lions, bobcats, and lynx, as well as snakes, birds of prey, alligators, and crocodiles.

HUMAN INTERACTION

Free-ranging and feral cats can be extremely damaging to wild bird or mammal populations in vulnerable or ecologically sensitive areas. Cats will kill chicks, young game birds, and waterfowl weighing up to 2 pounds. Less often they may attack larger chickens, game birds, ducks, and other waterfowl. They will also prey on domestic rabbits and their young.

Be very careful not to be bitten or scratched by a free-ranging or feral cat. According to the US Centers for Disease Control, cats are among the most common carriers of rabies and also transmit other diseases or parasites, including toxoplasmosis and hookworms, through their scat.

LEGALITIES

Legal status varies. Domestic cats are considered the property of their owners. Local ordinances may prohibit trapping a free-ranging cat and taking it to a shelter. In some areas, cats without ID are considered stray or feral and are under the control of the landowner.

RANGE

Domestic free-ranging, stray, and feral cats are found throughout North America.

Dealing with Feral or Stray Cats

Homes and Yards

- Do not feed feral or stray cats, because you will attract other pests and predators.

- Neuter and vaccinate outside barn cats, treat them for parasites, and feed them in secure areas. Be aware that your barn cats may kill young rabbits or poultry if they are not securely housed.

- Cover children's sandboxes when not in use.

- Do not place bird feeders in areas that provide brush, debris, or hiding places for feral or free-ranging cats.

- Use motion-activated sprinklers as a deterrent. Note, however, that cats quickly become habituated to most scare devices.

- Reduce or eliminate rodent populations in barns and outbuildings to discourage cats from hunting or taking up residence.

- Use dogs that are defensive against strange cats for the best protection around your home, yard, and barns.

- If it is legal in your area, live-trap roaming or feral cats to take to animal control agencies or shelters. The problem with "trap-neuter-return" projects is that when the cats return they continue to hunt, as even well-fed cats will.

Livestock Husbandry

Keep poultry and small animals such as rabbits in secure coops or pens. The same precautions used to exclude small predators from coops and pens will work with cats.

Fencing

- Cats can jump 6 feet straight up, and even higher if they can jump between areas or climb the fence material. Baffles around trees or posts can exclude cats from using them to climb.

- Mount electric scare wires 5 to 6 inches high and 4 inches outward from the top of the fence to prevent a cat from climbing over.

- Install mesh wire overhangs near the top of a fence.

- To exclude cats, the spaces in mesh fencing or between electric wires should not be greater than 2.5 inches.

To Bell or Not to Bell?

Belling a cat does not reduce its ability to hunt successfully. Cats use a slow stalk-and-ambush attack style for which a bell will not serve as a warning. In addition, even if the bell does ring, wild animals and birds do not associate the sound with an imminent attack.

DAMAGE ID: Domestic Cat

PREY ON

Chicks, young game birds, and waterfowl up to 2 pounds; larger chickens, game birds, ducks, and other waterfowl; domestic rabbits and their young

TIME OF DAY

Day or night

METHOD OF KILL

If a cat has caused damage to your birds or rabbits, it is very difficult to distinguish between free-ranging, stray, and feral cats. Pay attention to the time the attack occurred, as domestic cat attacks are more common during the day.

▸ Chicks, ducklings, and other small birds or rabbits often consumed entirely leaving only wings and scattered feathers or fur

▸ On larger animals, meaty portion of carcass consumed entirely

▸ Feathers (which may be found with loose skin still attached), feet, and wings not eaten

▸ Portions of carcass strewn over a large area, up to several square yards in size, with tooth marks left on exposed bones

▸ Well-fed domestic cats may leave the entire dead animal, consume only the head, eat all but the stomach or intestines, or carry the prey some distance away or to their home.

▸ Cats return to a successful hunting site on subsequent days.

TRACKS

Front 1–1⅝ inches long, ⅞–1¾ inches wide; rear 1⅛–1½ inches wide. No claw marks

 Front

 Rear

SCAT

Usually buried

3"

GAIT

Walk with hind print usually superimposed over front print

Birds of Prey

Accipitrimorphae

Raptors, or birds of prey, include the Accipitridae family of eagles, hawks, kites, harriers, Old World vultures, and the closely related *Cathartiformes* of New World vultures.

Found worldwide except Antarctica and some oceanic islands, Accipitridae hunt by sight during the day or dusk. They are small to large in size with sharp and strong hooked beaks, long wings highly adapted for soaring, strong legs, and clawed feet. They are also long-lived and monogamous.

The New World vultures include the black vulture, the turkey vulture, and the California condor in North America; the Central and South American vultures; and the Andean condor. Primarily scavengers, they locate carcasses by smell. These vultures are large birds with long, broad wings and a stiff tail used for soaring.

EAGLES

The only two eagles found in North America are the bald eagle and the golden eagle. Despite their similar appearance they are not closely related. Eagles have long been symbols of power, and these birds are indeed large and powerfully strong. The females are one-third larger than the males, an example of **reverse sexual size dimorphism.**

Eagles are daytime hunters; their eyesight is remarkable because the very large pupils provide acute vision at very long distances — from 4 to 8 times better than humans. It is believed eagles can spot small animals such as rabbits at a distance of 2 miles. They also have excellent color vision.

Both bald and golden eagles have heavy, hooked beaks and muscular legs. With wind assist, their powerful talons can carry prey weighing 15 pounds, though 5 to 8 pounds is more typical. If the prey is too heavy to lift, the bird may eat it at the site or dismember it and carry the pieces home.

Pairs are monogamous, building their **eyries**, or nests, high in trees or cliffs. The more dominant or older chick often behaves aggressively toward its weaker siblings. Parents may continue to provide for the surviving offspring for a few months after leaving the nest. If young eagles survive their first year, deaths are generally due to accidents with power lines or vehicles; loss of habitat; and illegal shooting, trapping, and poisoning.

Golden eagles are widely distributed throughout Europe, Asia, northern Africa, and North America.

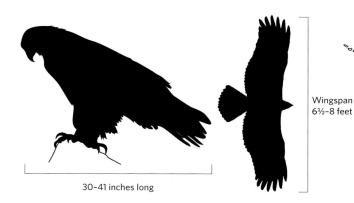

30–41 inches long

Wingspan 6½–8 feet

▲ Golden Eagle

▲ Goldens are most numerous in Alaska, and their range extends through the western areas of Canada and the United States, south into Texas. They are found in smaller numbers across the northern states and southern Canadian provinces, with scattered pairs elsewhere along the eastern seaboard and in southern states. Although most birds remain in their territories year-round, those in the very northern areas generally migrate to the southwestern United States. There are healthy populations of golden eagles in the West.

Golden Eagle
(*Aquila chrysaetos*)

With a 79-inch wingspan, golden eagles have broader wings than bald eagles do. The breadth and length of their wings afford them exceptional soaring abilities, with gliding speeds up to 120 mph, and full dives even faster. The birds average 30 inches long and weigh about 10 pounds. American golden eagles are generally smaller than Eurasian goldens.

Plumage and coloration. Color changes with age, with brown juveniles developing some white on their tail and wings. With maturity at 4 to 6 years, adult birds are dark brown, with some grayish brown on the inner wing and tail, and the characteristic golden brown or bronze head and nape. The lower legs are feathered to the toes, as with other "booted" eagles. The bill is dark horn colored, and the feet are yellow.

HABITAT AND BEHAVIOR

Often seen over open coastal areas, foothills, high mountains, and bluffs or cliffs, above valleys where they hunt prey in brushy lands and grassy plains, golden eagles maintain a large hunting territory of up to 35 square miles or more. They will engage in confrontations with other raptors. On flatter prairies, they live around steep wooded river valleys, and some birds will migrate to wetlands and coastal estuaries.

Hunting and foraging. Rabbits make up 70 percent of the golden eagle's diet. Although it is uncommon, they can kill prey weighing up to 65 pounds, such as deer, antelope, bighorn sheep, and even predators such as coyotes, bobcat, and foxes. Golden eagles will more commonly take lambs, kids, calves, and stock weighing up to 50 pounds. They were historically used for falconry in the Old World, even hunting large animals such as wolves.

Although more active and bolder than bald eagles, goldens spend most of their day perching. Sometimes they hunt in pairs to exhaust their prey. They do not tend to cache food. They primarily hunt small mammals, birds, and reptiles, but they will also eat carrion.

LIFE CYCLE

Breeding season varies with geography. Golden eagles often build several large nests in their territory, primarily on cliffs although they may use trees or manmade towers. They will alternate between nests from year to year. The female lays 1 to 4 eggs and performs most of the incubation, although both partners provide for the nestlings. The golden eagle life span is 20 to 30 years.

Bald Eagle
(*Haliaeetus leucocephalus*)

Bald eagles are about 31 inches long, with an 80-inch wingspan, and weigh 9.5 pounds, although they can be larger. Birds in the southern range are smaller, and Alaskan birds are larger.

Plumage and coloration. Color changes occur as the birds grow. Juveniles lack the distinctive white patches and are easily confused with adult golden eagles. Maturing at 4 to 5 years, the dark brown adults have the distinctive white head and tail. The beak and feet are yellow, the lower legs are bare of feathers, and the tail is long and moderately wedge-shaped.

HABITAT AND BEHAVIOR

As a sea eagle, the bald eagle prefers old-growth woods, forests, and mountain ridges near seacoasts, large lakes, rivers, and southern wetlands — although it is found in western deserts — and tends to avoid human areas. Nesting and hunting territories are much smaller than those of the golden, especially if near water. After nesting, bald eagles often group together near good food sources.

They will remain in northern areas wherever the water remains open, but they also migrate to the southwest. Bald eagles can also move around the country during the summer. Birds that nest very early in Florida will travel up into the Northeast and Canada during the summer, returning in the

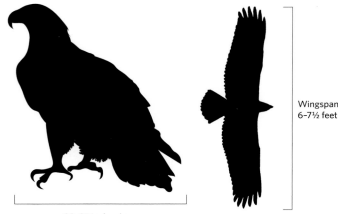

Wingspan
6–7½ feet

30–37 inches long

▲ Bald Eagle

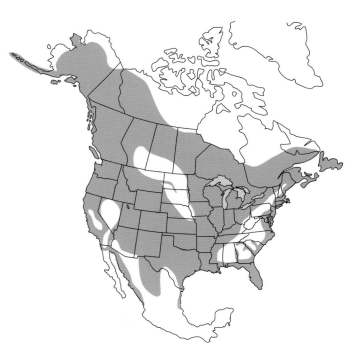

▲ Bald eagles are found throughout most of the United States and Canada, from northern Alaska to Newfoundland (excluding arctic areas), and south from Florida to California. The breeding population in Alaska and Canada is larger and healthier than farther south in the lower 48 states, where the numbers of breeding pairs is estimated at only a few thousand. This is a significant recovery, however, from only 400 pairs in the 1950s. The largest populations are found in Alaska, Florida, the Pacific Northwest, and the upper Midwest.

▲ The United States was originally home to as many as 500,000 nesting pairs of bald eagles. Today about 5,000 pairs live in the lower 48 states.

fall. Birds from the Great Lakes may move to the Atlantic coast, while northeastern birds can travel to the Appalachians.

LIFE CYCLE

Bald eagles nest very high in trees, cliffs, towers, or poles. Their eyries are larger than any other North American bird's, are often reused for several years, and are heavily defended. Both parents incubate 1 to 3 eggs and care for the nestlings, which are vulnerable to many predators. Birds mature at age 5, with a life span of 20 to 30 years if they survive their first year.

Hunting and foraging. Powerful flyers, bald eagles primarily eat fish, although they will take large waterfowl and other birds, small mammals, and reptiles; occasionally, they prey on deer and antelope. They will also scavenge carcasses and steal prey from other predators. They often perch, awaiting their prey, and then swoop down and snatch fish near the water surface. They may also harass waterfowl until the smaller birds are exhausted. Heavy fish or birds may be dragged over the water to shore. Bald eagles can store food in their crop for several days. Some prey on lambs and kids, calves, or adult animals, but this occurs much less often than with golden eagles.

Dealing with Eagles
Homes and Yards

Attractants include rodents and free-ranging birds and rabbits.

Livestock Husbandry

Bald eagles can be attracted in large numbers to large-scale pastured poultry operations. Newborns are very vulnerable during birthing seasons. Eagles will also prey on 3- to 4-week-old lambs and kids as they begin to wander or play farther from mothers, but not generally after 6 weeks of age.

- Set up birthing sheds or protected pens less than 1 or 2 acres in size. Temporary pens for night and early morning are very useful since eagles are unlikely to enter smaller spaces.

- Use livestock guardians, particularly dogs that alert to aerial attacks. Livestock guardians are more effective in areas small enough for them to actively protect, rather than very large spaces with scattered stock or poultry.

- Provide brushy grazing or shelters, which is safer than large open pastures.

- Provide an active herding and human presence to discourage eagles.

- Remove all carrion. Do not feed eagles.

- Mount or suspend clothed scarecrows, with movable arms, on high points near nightly bedding areas, to scare eagles for up to 3 weeks (after which they become used to them).

- Use portable electric netting to subdivide large pastured poultry areas.

- Use netting or wires over pastured poultry to disrupt aerial attacks.

- Use rounded corners to reduce smothering during an attack.

- Remove roosting sites.

HUMAN INTERACTION

Eagles are important apex predators in the ecosystem. Golden eagles are more likely to prey on young lambs, goats, and calves than bald eagles are. Occasionally eagles will prey on young swine or other animals such as domestic rabbits, waterfowl, poultry, and small pets. Primary losses of larger stock occur in grazing areas in Colorado, Wyoming, Montana, Utah, and Texas but also happen elsewhere in their range. Predation of lambs and kids on range during birthing seasons appears to be increasing, but both eagles feed on livestock carcasses left by other predators, which can confuse observers.

LEGALITIES

The federal Bald and Golden Eagle Protection Act and Migratory Bird Treaty Act protect bald and golden eagles, their nests, and nest sites. Killing eagles without permit is illegal, as is possessing, selling, or trading any live or dead eagle, feathers, or eggs. Harassing, harming, hunting, shooting, and trapping are also illegal. Hazing with gunfire, explosives, and airplanes is prohibited without permit. Only USDA-APHIS-ADA personnel are permitted to conduct permitted depredation activities after a formal assessment. Relocation is generally a failure, with most eagles returning to their home territories.

DAMAGE ID: **Eagles**

PREY ON

Rabbits, lambs, kids, calves, piglets, domestic rabbits, waterfowl, poultry, and small pets. Goldens will prey on lambs, kids, calves, and stock up to 50 pounds.

TIME OF DAY

Daytime, often near dawn

METHOD OF KILL

If a carcass is still present, determine whether eagles were opportunistically feeding on a stillborn or dead animal. Examine the carcass for the signs of an eagle attack.

▸ Small animals sometimes taken away

▸ Talon punctures at the top of the skull, neck, or body

▸ Talon punctures, usually deeper than tooth punctures and less crushing than large predator bites, although compression fractures or bruising may be present. Rather than 4 canine tooth punctures, the 3 front talons leave marks 1–2 inches apart, with one rear toe 4–6 behind those. On small lambs or kids, all 4 talon punctures may not be found.

▸ On larger animals, multiple talon wounds in the back or upper ribs, with death caused by internal hemorrhage or lung collapse

▸ Carcass skinned out, leaving most of the skeleton intact, although an eagle may eat the lower jaw, nose, ears, palate, and brains, as well as the ribs on very young animals

▸ Skin cleaned of blood and flesh

▸ Rumen not eaten in larger animals

▸ Eagle feces around the kill; tracks visible in the dirt

TRACK

Classic, about 6–7 inches long and 4½ inches wide. Talon marks visible.

GAIT

Walking stride about 18 inches

Golden Eagle

Bald Eagle

SCAT AND PELLET

Semiliquid, white and brown

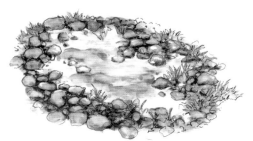

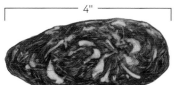

Golden Eagle Pellet

Bald Eagle Pellet

191

HAWKS

awks are a group of small to medium-sized raptors that hunt during the day. In North America the many hawk species all have sharp, forward-facing eyes; hooked bills; and talons. The females are generally larger than the males.

These birds' many hunting techniques include watchful perching; low **coursing** over open areas; **hovering** by flapping their wings; **kiting**, or hanging in the wind or thermals; level-flying pursuits; **stooping**, or diving from high; and **flycatching**, or grabbing with talons in midair.

Both the Cooper's hawk and the Northern goshawk are members of the *Accipiter* genus, while the red-shouldered and red-tailed are buteos. More rarely seen, accipiter hawks are forest dwellers with longer tails and shorter, rounder wings than buteos. They hunt from hidden perches and are seen only infrequently, flying with stiff, flapping wing beats followed by a glide. These hawks are known for their sudden, fast, horizontal pursuits from a hidden perch.

More often seen, the *Buteo* hawks are more broad-winged and sturdy. They are commonly observed as they soar over open areas. These hawks are more likely to make rapid dives and fierce pounces onto prey. In other areas of the world, *Buteo* hawks are called **buzzards**, a name that was applied to vultures in North America.

Cooper's Hawk
(*Accipter cooperii*)

DESCRIPTION

The Cooper's hawk has a long, lean body shape and is about 16 inches long with a 31-inch wingspan, weighing up to 1 pound or so. Those found in the western areas of their range are 20 percent smaller than birds in the East. The Cooper's has a dark gray-black cap, or **crown**; a lighter color on the back of the neck, or **nape**; a body tinted slate blue-gray above with reddish bars on light-colored underparts; and wide, dark bands on the tail tipped in white. It can be hard to distinguish from its close relative the sharp-shinned hawk, which is considerably smaller at 11 inches long with a 23-inch wingspan. The sharp-shinned also has a more extensive range in Canada than the Cooper's hawk does.

HABITAT AND BEHAVIOR

This hawk lives in deep forests, more open and scattered woodlands, semiarid areas, and wooded suburban and urban areas. Many Cooper's hawks remain as year-round residents throughout their range, although birds in more northern areas will migrate south. A swift, agile flyer, it usually hides on a perch before swooping down on prey. Cooper's hawks can also fly low to the ground, maneuvering quickly over obstacles to snatch prey, or they can chase prey by flying and running along the ground. They pursue birds or small mammals and occasionally eat amphibians or reptiles.

LIFE CYCLE

Many Cooper's hawks mate for life. The breeding season can begin in March, with the male choosing a nest site and the female laying 3 to 6 eggs. The female incubates the eggs while her mate provides for her. After the eggs hatch, both parents hunt for the nestlings. Reaching maturity at age 2, Cooper's hawks can live up to 12 years. They are vulnerable to both owls and other hawks.

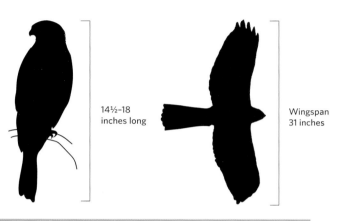

14½–18 inches long

Wingspan 31 inches

▲ Cooper's Hawk

▲ Cooper's hawks are often spotted hunting for birds at backyard feeders.

▲ Cooper's hawks are common throughout the continental United States and southern Canada.

Northern Goshawk

(Accipiter gentilis)

DESCRIPTION

Larger than the Cooper's hawk, the Northern goshawk is also bulkier and stronger; average-sized birds are 21 inches long, with a wingspan of 41 inches, and weigh 2 pounds. A white eyebrow separates the black crown and cheek patches; the back is dark to slate gray with lighter white-gray underparts; the wings are two-toned in gray from above and below; and the tail is lightly barred.

HABITAT AND BEHAVIOR

In North America, the goshawk is known to be a more secretive bird, living away from human areas in mature forests, including mountainous regions. Goshawks in northern areas will migrate to warmer or lower areas; however, most birds remain in their chosen nesting territories. They seek out tall, mature trees for nesting, with access to smaller, more open areas for hunting.

Northern goshawks hunt at the edges of their wooded territories. This hawk is a bold and powerful predator able to lift relatively large prey such as foxes or raccoons, in addition to a regular diet of rabbits, squirrels, birds, and reptiles. At times, they cache their prey in trees.

Northern goshawks mate for life. The breeding season begins in early April, with the pair repairing their old nest or building a new one and aggressively defending their territory. The female lays 2 to 4 eggs, incubating the eggs and caring for the nestlings while the male provides for them all. The young reach maturity at age 1 and are believed to live to 11 years or longer. They are vulnerable to owls and other hawks.

The Northern goshawk is expanding its range southward in both eastern and western states, especially during fall and winter when they are in search of prey.

18-27 inches long

Wingspan
3-4 feet

▲ Northern Goshawk

▲ Northern goshawks are found from central Alaska, the Yukon and Northwest Territories south across Canada, the northern United States, and the mountainous areas of the western states. They very rarely appear in the southeastern states.

Red-Tailed Hawk
(Buteo jamaicensis)

DESCRIPTION

Red-tailed hawks average 22 inches long with a 48-inch wingspan and weigh 2.5 pounds; larger birds are found in the northern areas. The wings are broad and wide. With many subspecies, red-tailed hawk coloration is variable with the region — from light auburn or pale brown to red or dark rich brown above and lighter colored underparts, usually with a broad dark striped band. Most adults display the characteristic short and wide red tail, with a redder cinnamon color on top and pale below.

Often observed soaring over open areas, the red–tailed is the most familiar hawk in North America.

HABITAT AND BEHAVIOR

The most commonly seen hawk in North America, red-tails are highly adaptable, capable of living in forests, plains, deserts, coastal and inland wetlands, pastures and agricultural fields, and very urban settings. They prefer mixed areas with high trees, utility poles, or other areas for perching. Northern birds migrate south, while others remain year-round.

Red-tails also spend much of the day soaring and hovering or kiting over their territory, both defending their space and hunting. They can be identified by their frequent high-pitched, raspy scream. They are capable of steep and very fast dives of 120 mph. Red-tails tend to swoop from a perch to grasp prey, snatch birds in flight, or pursue them on the ground. With powerful talons to hunt and grasp prey, they feed primarily on rodents but also on snakes, lizards, small mammals, occasionally birds, and even carrion. They will also aggressively steal prey from other large hawks or eagles. In cities, they prey on pigeons, birds, or rats. They are known to opportunistically and boldly prey on domestic pets and poultry. They do not cache food.

Monogamous pairs use the same nest for several years in a tall tree, cactus, cliff, or building ledge. They defend their territory. Females lay 1 to 5 eggs. While the pair both incubate the eggs, the female stays with the nestlings while the male hunts to feed everyone. The young reach maturity at age 3. With few predators, red-tailed hawks live 20 to 30 years.

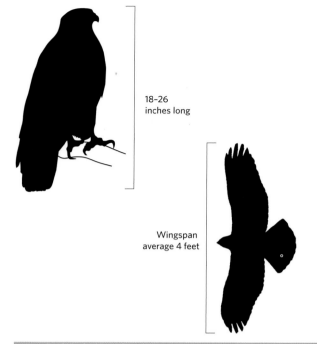

18-26 inches long

Wingspan average 4 feet

▲ Red-Tailed Hawk

▲ Red-tailed hawks are found in the entire continental United States, and most of Canada, and Alaska south of the tundra.

Red-Shouldered Hawk
(*Buteo lineatus*)

DESCRIPTION

Smaller and more compact than the red-tailed hawk, the red-shouldered hawk has similar broad, rounded wings and a wide medium-length tail. About 20 inches long, with a 40-inch average wingspan, adults weigh about 1.5 pounds. There are separate Eastern and Florida types, as well as one in California. Adults in the East have a brown head; checkered markings of dark brown and white above; a reddish peach breast and bars on light-colored underparts; and a black-and-white banded tail. Florida birds are paler in color. The California type has a richer orange breast and underside of the wings.

HABITAT

Red-shouldered hawks prefer mature forests near water, such as rivers or forested swamps, where they perch in trees or on wires. They are increasingly seen in more urban environments. They live year-round in many areas of their range, except in the northeastern United States and southern Canada, from where the hawks migrate to northern Mexico.

They hunt by either perching in a tree or soaring and dropping down on their prey. They primarily eat rodents and small mammals, as well as some snakes, lizards, amphibians, small birds, crayfish, and even large insects. Near farm fields, they prey on rodents but also on poultry. They are known to cache food in trees.

These hawks are territorial, using clear whistled calls to mark their claims.

LIFE CYCLE

Red-shouldered hawks are monogamous and usually reuse their nest. The female lays 3 to 5 eggs, which both parents incubate. Later the male feeds his mate and nestlings. The young reach maturity at age 1 and are known to live almost 20 years. In the nest, both adults and nestlings are vulnerable to owls and raccoons.

HUMAN INTERACTION

Hawks play an important role in keeping down the populations of small pests and rodents, but will also opportunistically prey on poultry, game birds, animals such as rabbits, or even small pets. The Northern goshawks, although more uncommon, are bolder in preying on any free-range poultry, including larger birds such as geese and turkeys.

▲ In its eastern territory, the red-shouldered hawk is seen less frequently than in the past.

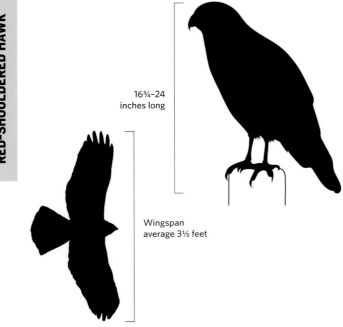

16¾–24
inches long

Wingspan
average 3⅓ feet

▲ Red-Shouldered Hawk

▲ Red-shouldered hawks are found in the eastern United States and southeastern Canada, and in a separate population in western California and coastal areas north through British Columbia. They are Threatened or Endangered in several states.

Northern goshawks have long been used in falconry; red-tailed hawks are also popular with falconers in North America.

LEGALITIES

The US Migratory Bird Treaty Act and other federal and state laws protect hawks. They cannot be harassed, captured, harmed, or killed without a permit. Some hawks may also be listed as endangered in some states. Consult local USDA-APHIS Wildlife services, the US Fish and Wildlife Service, or your state wildlife service for specific regulations.

Dealing with Hawks
Homes and Yards

- Remove attractants, including bird feeders or animal feeding stations.

- House small dogs, cats, and pet birds in covered pens or kennels if left outside.

Livestock Husbandry

- Cover coops, runs, and pens well.

- Cover larger yards with grid systems of wire, reflective tape, Kevlar cord, or netting.

- Do not locate free-range areas near potential perching sites such as trees or poles. These areas should be provided with protective shelters. Outfit tall posts with electric pole shockers, metal cones, or spikes to discourage perching.

- Keep a rooster. Roosters often give useful alarms to their flock, providing time to hide in protective shelters or brush.

- Livestock guardian dogs (LGDs) will alert, bark, and charge at raptors.

- Use noisemakers, whistles, pyrotechnics, and gunfire to scare off hawks, until they become habituated.

DAMAGE ID: **Hawks**

Hawks usually kill only one bird per day, although they may return to eat more the next day or take another bird.

PREY ON

Free-range poultry, waterfowl, turkeys, pigeons, game birds, rabbits

TIME OF DAY

Generally daytime, but also dusk or dawn

METHOD OF KILL

- Entire bird or animal missing
- Feathers left at site of attack
- A heavy carcass left nearby; bloody puncture wounds on the back and neck or breast; the head and crop eaten first
- Hawks will pluck feathers before eating. If feather base is smooth and clean, the bird was plucked soon after the kill. Beak marks may also be found on the shafts.
- If the feathers have a small amount of tissue attached to the base, they were pulled from a dead bird already cold, meaning the hawks were scavenging carrion and did not kill the bird.

TRACK

Red-tailed hawk: Classic, 3¾–5⅜ inches long, 2½–4 inches wide

Northern goshawk: Classic, 4½–4¾ inches long, 2½–3 inches wide

GAIT

Walking stride: 2¾–5½ inches
Running stride: 12–15 inches

Cooper's Hawk

Red-Tailed Hawk

Red-Shouldered Hawk

Northern Goshawk

SCAT AND PELLET

Semiliquid, white and brown. Hawks often defecate at their kill site.

Cooper's Hawk Pellet

Red-Shouldered Hawk Pellet

Red-Tailed Hawk Pellet

VULTURES

The two New World vultures found in the United States are the black vulture (*Coragyps atratus*) and the turkey vulture (*Cathartes aura*). Commonly called buzzards, the two are often confused, but their diet is very different. While turkey vultures specialize in finding and eating carrion, black vultures are also predatory and responsible for the death of pets, poultry, and farm animals.

The black vulture's range is extending northward, where it roosts near cities and towns.

Black Vulture
(*Carthartes aura*)

DESCRIPTION

Both adult and juvenile black vultures have glossy black bodies with small, dark gray, bare, wrinkled heads and strongly hooked bills. The tail feathers are short and the undersides of the wings are dark gray to black with a white patch at the end of each wing. The legs are grayish white. The black vulture is 22 to 29 inches long, with a wingspan somewhat less than 5 feet, and weighs about 4 pounds.

In contrast, the red- or pink-headed turkey vultures are lankier in appearance, with longer tails and wings that are gray underneath instead of black. Juvenile turkey vultures have blackish heads.

HABITAT AND BEHAVIOR

Black vultures live in long-term family relationship groups and gather socially in roosts, which can include more than 1,000 individuals. Typically roosting in lightly forested areas near open, warm areas such as fields or pastures where they feed, black vultures are also found near garbage dumps and Dumpster sites. Roosts are located in trees or on transmission towers, where the birds spend the night and the early morning waiting for the air to warm enough for thermals to develop and support them as they forage later in the day.

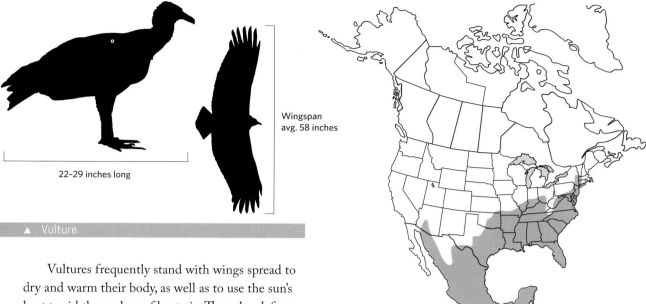

▲ Vulture

Vultures frequently stand with wings spread to dry and warm their body, as well as to use the sun's heat to rid themselves of bacteria. They also defecate on their own legs to cool the blood vessels, leaving white streaks. Neither the feet nor the talons are well adapted to grasping. Vultures grunt or hiss when disturbed.

The two species can be distinguished by their flying style. Black vultures must flap frequently and at low altitudes they fly in a labored, shallow, choppy rhythm. They can glide for only a short distance unless there are strong winds, although at higher altitudes they glide on thermal wind currents. Turkey vultures are more capable of gliding at low altitudes. Black and turkey vultures are often found soaring together with hawks during the day.

LIFE CYCLE

Although black vultures bond for life, they generally do not reproduce until about age 8 or later. Breeding goes from January through March, depending on location. Black vultures lay 1 to 3 eggs in hollow trees or stumps, on flat surfaces under vegetation, in shallow caves, or in abandoned buildings. Although no nesting materials are used, vultures occasionally decorate the area around the nest with shiny or bright-colored shards of plastic, glass, or metal. The pair continue to feed their offspring for many

▲ Black vultures are found from the southeastern Atlantic coast north to southern areas of Connecticut, New York, Pennsylvania; west to Ohio, Indiana, Illinois, and Missouri; and south to eastern and southern Texas and south-central Arizona. More recently, black vultures have been observed farther north and east, and in California. Most depredations are reported in Virginia, Texas, South Carolina, Florida, and Tennessee.

Over the past century, turkey vultures have extended their summer breeding range throughout most of the continental United States and into areas of southern Canada.

months after hatching. Groups of turkey vultures will share roosts with black vultures; hybrid chicks have occurred in captivity. In captivity, vultures have lived 20 years.

Unlike the more solitary turkey vulture, black vultures feed in groups. Visually oriented, black vultures have a poor sense of smell and often move in on carrion located by turkey vultures and then bully them away. They will also eat very ripe or decaying fruit or vegetables. Although eagles may kill vultures in conflicts, adult vultures do not experience predation.

HUMAN INTERACTION

All vultures perform the valuable job of eating carrion. Black vultures also prey on eggs, young poultry, and waterfowl; young calves, lambs, kids, and piglets; and weakened adult animals, especially cattle. Both species are attracted to afterbirths, stillborns, and the fresh droppings of calves. Vulture roosts can cause an accumulation of guano (droppings) underneath.

LEGALITIES

Vultures are protected by the US Fish and Wildlife Service. Landowners must have a Migratory Bird Depredation permit to remove them.

Dealing with Black Vultures
Homes and Yards

- Remove dead, ill, or weakened animals, and nearby roadkill.

- Place birthing animals in protective buildings, sheds, or closely monitored pastures or paddocks.

- Discourage vultures with loud sounds such as pyrotechnics and gunfire.

- Hang a vulture carcass or a taxidermy effigy of a dead vulture in the roost location to disperse the roost permanently.

DAMAGE ID: Black Vulture

It is necessary to evaluate through a field autopsy whether dead newborns were stillborn or killed by black vultures after birth. A cow may have also died while giving birth or soon afterward.

PREY ON

Eggs, young poultry and waterfowl, young calves, lambs, kids, piglets, weakened adult animals, especially cattle

TIME OF DAY

Daytime, once the air has warmed up and thermals have developed

METHOD OF KILL

▸ Predation by a group of 8–40 birds

▸ Calves and other young animals attacked as they are being born or soon after when the mother is tired or weak

▸ Older calves and lambs surrounded, with eyes, nose, tongue, rectum, and genitals attacked

TRACK

Classic, with 3 distinct toes, 4½ inches long.

SCAT

White fluid

CHAPTER 13

True Owls

Strigidae

The large family of "true" owls is found worldwide except for Antarctica and a few islands. Different species vary greatly in size, but all have upright bodies; large, broad heads; large, long, broad wings; short tails; and sharp talons.

Owls have a variety of traits that help them hunt, including excellent hearing and eyesight. Their large eyes have extraordinary night vision, enabling them to find prey and fly at night. Binocular vision allows them to judge the height, weight, and depth of their quarry. They also have **binaural** or directional hearing, giving them exceptional sound localization and enabling them to hunt successfully at night. They have the ability to rotate their heads left to right and nearly upside down, as their forward-facing eyes do not move in their sockets.

Generally solitary except during breeding and nesting, both great horned and barred owls do not migrate but do establish changeable territories. Mating season varies with geography, from January through April. Essentially nocturnal, owls spend the daytime perching in concealed foliage or nesting, at times disrupted by smaller birds that attempt to drive them away. They rarely hunt during the day, except on dark days or to feed nestlings.

Great Horned Owl
(*Bubo virginianus*)

Often seen or heard, the large great horned owls are found throughout most of North America. Long the symbol of wise intelligence, this image of a silent, calm bird perched on a tree branch or fence post is familiar to all.

DESCRIPTION

Great horned owls have thick yet lightweight bodies, weighing about 3 pounds, with wide wings capable of near-silent flight. They are generally about 20 to 22 inches tall with a 44- to 48-inch wingspan, and females are larger than males. This owl has a round face with large, forward-facing, yellow eyes and feather tufts on top of its head, which lie flat in flight. Color varies by regional adaptation, as do size and behavior. Camouflage is provided by mottled gray, black, and reddish brown colorations with dark and white bars on the underside and a white spot on the throat.

HABITAT AND BEHAVIOR

Although found in dense forests, great horned owls prefer areas with more open woodlands, fields, and pastures. They also inhabit deserts, swamps, wetlands, and even urban environments. Males give a series of 4 to 5 deep hoots to mark their territory, often in early evening or before dawn, and to locate mates, while females hoot only during mating season. Owls may also use a loud screeching sound as they attack prey.

Hunting and foraging. Great horned owls hunt during the night, primarily for small mammals such as rabbits, skunks, rodents, and birds, including poultry, but also eating reptiles, amphibians, fish, and insects. From a perch, the owl swoops down to grasp its prey with talons, either in the air or off the ground.

Wingspan 3⅔–4 feet

20–22 inches long

▲ Great Horned Owl

▲ Great horned owls are found throughout the entire continental United States and major portions of Alaska and Canada, excluding arctic areas.

▲ The great horned is the most common owl of both North and South America. It has yellow eyes and distinctive ear tufts.

LIFE CYCLE

The mated pair share incubation duties, chick rearing, and protection of the territory around their nest. The number of eggs laid depends on the availability of food. Males hunt more than females during parenting time. Males choose a new nest every year, usually the abandoned nests of birds or even squirrels, but they may also nest on cliff faces, in barn lofts and silos, and on transmission towers; in some areas they will use brushy areas or even the ground.

Great horned owls are apex predators. Deaths result from conflicts with other owls, accidents or hunting injuries, and predation on their nests. They mature in 1 to 3 years, and they live well into their late twenties or longer.

Barred Owl
(*Strix varia*)

Similar in size to the great horned owl, the barred owl was originally found in the forested eastern states but has now expanded its range far beyond the Mississippi River, north through Canada into southern Alaska, and south into Washington, Oregon, northern Idaho, western Montana, and northern California. The treeless Great Plains served as a barrier to its spread until settlers altered the habitat through tree planting and fire suppression. Beginning in the late 19th century, the barred owl and other bird species made use of these trees and the forested corridors of the major rivers as they moved west and northward. Barred owls reached Washington State by the mid-1960s, and were in California 10 years later.

This expansion brought them into the northern portion of the habitat of a similar wood owl, the rare spotted owl. Where these two owls share the same area, the more aggressive barred owls are more successful competitors for the same prey, which drives down the spotted owl population. They occasionally interbreed, giving rise to a hybrid owl known as the **sparred** owl, which looks much like the highly endangered spotted owl. There have been controversial proposals to remove barred owls from the spotted owl range.

DESCRIPTION

Large and stocky, the barred owl weighs 1 to 2.3 pounds, ranges from 16 to 25 inches long, and has a 38- to 49-inch wingspan. It differs from the great horned in that it lacks ear tufts and has dark brown, nearly black eyes and a rounded tail.

Plumage and coloration. The barred owl's upper parts are mottled gray-brown, while its chest has distinctive vertical stripes. Its face is pale gray, and its legs and feet are covered with feathers.

HABITAT AND BEHAVIOR

These owls prefer old, mature, dense mixed forests and wooded river bottoms and swamps, although they have moved into dense coniferous forests and nearby logged areas in the Northwest. They are now less common in areas of the South due to the loss of habitat. On the other hand, they have adapted to life in the suburbs, where they are reproducing faster than in their native habitat. The presence of a great horned owl will discourage the less aggressive barred owl.

During the day, barred owls roost in trees although they may call or hoot to other owls. Nesting in tree cavities or the abandoned nests of hawks, crows, or squirrels, they often return to the same spot each year. The female tends to brood the eggs and remain with nestlings, while her mate hunts for both her and their young.

Although nocturnal, barred owls usually hunt at dawn or dusk, but they will hunt during the day to feed nestlings or if the sky is dark. Waiting on a high perch or flying low through trees, the barred owl swoops down to grasp prey with its talons. Rodents and small mammals are their primary food, and occasionally birds. Barred owls walk into the water to take amphibians, fish, and crustaceans. They also feed on large insects attracted to lights.

Barred owls live up to 10 years. In suburban areas, the main cause of death is a car accident. In the wild, great horned owls and cats take young birds.

▲ The barred owl, slightly smaller than the great horned owl, has dark eyes and no ear tufts.

Also Known as the Hoot Owl

The barred owl is often called the *hoot owl* for its clear series of 8 hoots, ending with a long descending note. To identify the call, birders use the mnemonic "Who cooks for you, who cooks for you all."

16–25 inches long

Wingspan 38–49 inches

▲ Barred Owl

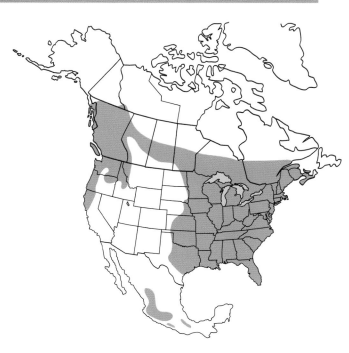

▲ The barred owl is found in eastern and central states north through Canada into southern Alaska, and south into Washington, Oregon, northern Idaho, western Montana, and northern California.

Other Owls

Many other owls are residents of the United States and Canada. They are less common predators of poultry or small farm animals due to their small size, habitat, or habits. Although not common near farmyards, there are two other large owls capable of preying on small animals and poultry.

The heaviest owl, the snowy, is a northern resident that nests in the tundra but winters farther south in Canada. In recent years, snowys have been spotted in various locations in the United States during the winter. Active day and night, they primarily hunt lemmings and small rodents, occasionally preying on small mammals such as hares and water and game birds.

The largest owl species, the great gray, is primarily found in the boreal forests and mountains in Canada and Alaska, and northernmost areas of the Rocky Mountains. The Sierra Nevada in California is home to a very small population of about 100 owls. The great gray has a very long, mottled gray body; a broad gray face with a white beard; and yellow eyes. It preys primarily on rodents.

Medium-sized barn owls are frequently seen around human structures in coastal areas and the central and southern states. Avid hunters of rodents, barn owls differ from the great horned by their heart-shaped face, dark eyes, gold-buff color, and lack of ear tufts. Barn owls belong to the related family Tytonidae.

HUMAN INTERACTION

Important hunters of rodents around farms, owls can also prey on poultry, rabbits, game birds, and water-fowl. At times, there are reports of puppies, kittens, and adult cats lost to owls. While the great horned owl is most often linked to poultry losses, the barred owl is increasing its range and is now seen in suburban areas, where it has been observed carrying away cats and backyard poultry. Other owls do occasionally prey on poultry as well.

LEGALITIES

Owls are protected under the Migratory Bird Treaty, and it is illegal to kill one without a permit.

Dealing with Owls
Homes and Yards

Unattended kittens, puppies, cats, pet birds, rabbits, and poultry are vulnerable. Provide covered housing for animals left outside at night.

Livestock Husbandry

The most effective protection is to secure poultry or rabbits at night in a predator-proof coop or covered pen. Owls will fly or walk on the ground to enter coops. They will also swoop through openings in runs or coops.

- Cover the top of the pen or coop with hardware cloth or tight mesh. Install wires or reflective tape to disrupt flight paths and areas of potential attack.

- For larger yards, set up grid systems of wire, monofilament, or Kevlar cord, covered with netting.

- Seal or cover all access around gates.

- Provide nighttime hiding places for free-ranging birds, waterfowl, and rabbits.

- Add a rooster to your flock, as a warning system.

- Avoid free-ranging sites with high perches (such as trees), although owls will also use fence posts.

- Use livestock guardian dogs if socialized to poultry for patrolling around poultry areas.

- Set up scare devices such as noisemakers, motion-activated sprinklers, lights, scarecrows, and reflective items. However, owls do become habituated to routines, lights, and many scare devices. Barred owls, in particular, are attracted to lights to hunt for large insects.

Bird Tracks

Bird gaits include walking, running, hopping, and skipping, and the tracks come in two different types, depending on the species.

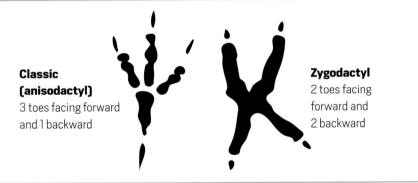

Classic (anisodactyl)
3 toes facing forward and 1 backward

Zygodactyl
2 toes facing forward and 2 backward

DAMAGE ID: Owls

PREY ON

Poultry, rabbits, game birds, waterfowl

TIME OF DAY

Night

METHOD OF KILL

- ▶ Prey usually eaten on the spot, leaving lots of feathers, but sometimes carried away to eat elsewhere
- ▶ Head and neck eaten or removed

- ▶ Larger prey taken to another perch and torn apart
- ▶ Another bird taken the following night
- ▶ Droppings and pellets beneath where the owl perched

TRACK

Zygodactyl: 2 toes pointing forward and 2 backward. Owls can rotate the 4th rear toe in different positions including out to the side. Size varies with species. Talon mark visible.

SCAT AND PELLET

Semiliquid, white

CHAPTER 14

Crows

Corvidae

American crows, ravens, magpies, and jays belong to the large crow family, found around the world. Medium to large in size, the corvids have strong bills and feet, large wingspans, and nostrils covered in bristle-like feathers. Male and female are much alike in appearance and size.

Mated pairs will noisily and aggressively defend a nesting territory, even attacking larger animals or humans. Corvids also mob together to chase away larger predator birds. Most corvids do not migrate, although they will move in search of food or away from very cold winters. During the winter, they often gather in large flocks to forage or share communal roosts.

Omnivorous, corvids will eat almost anything they discover while walking on the ground or branches. They hunt small mammals, nestling and small birds, amphibians, crustaceans, and insects. They also gather berries, fruit, and seeds. They are important scavengers of carrion, which has led them to associate closely with garbage dumps, animal feed, and other human-provided food sources. Corvids will also hide or cache food in multiple locations, returning months later to the correct spot. After eating, corvids clean their beaks on vegetation or dirt.

Known for their intelligence and ingenuity, corvids use tools and live in complex social groups. They are very vocal birds and, in addition to making hoarse caws or cries, alarm, and comfort calls, they mimic other animal sounds or noises. Even though people keep ravens or crows as pets, these birds are widely viewed as pests. Populations have increased, as corvids are well adapted to human activity.

American Crow
(*Corvus brachyrhynchos*)

DESCRIPTION

American crows average 17.5 inches in length, with a 39-inch wingspan, and weigh around 1 pound. Their wings are broad, and they have a short, fan-shaped tail. There are 4 regional subspecies, which differ slightly in size and proportion. A slightly smaller relative, the Northwestern crow, is found in the coastal areas of western Canada and Alaska.

Plumage and coloration. The feathers are iridescent black, and the bill, feet, and legs are also black. Occasionally there can be some white patches on the wings.

HABITAT AND BEHAVIOR

Adaptable to most habitats, crows prefer lower elevations in wooded areas near open areas such as grassland, cropland, or shoreland. Some Canadian and American crows migrate southward, but not long distances; in most areas, however, they are full-time residents. They are often seen in small groups foraging on the ground. Although they do scavenge carrion, they are active hunters. While they will feed on crops, they also eat many insect pests.

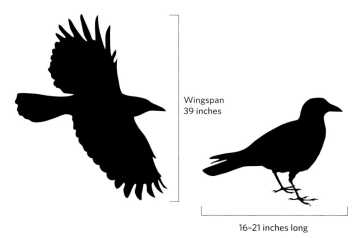

Wingspan
39 inches

16–21 inches long

▲ American Crow

Intelligent and highly observant, the American crow has adapted easily to life in cities and suburbs.

LIFE CYCLE

Crows are monogamous and occupy the same territory each year, forming cooperative family breeding groups of up to 15 birds. The yearling offspring remain with their parents, helping to care for nestlings and defend territory. Crows prefer to nest in trees but will use shrubs or even ledges on buildings. Beginning in early April, the female lays 3 to 6 eggs. Although they mature at age 2, young birds usually do not establish their own nests until 4 to 5 years of age. The life span averages 7 to 8 years although captive birds live much longer. Raptors and owls prey on adult birds.

During the nesting season, small groups of nonbreeding birds can also form small floater flocks. When nesting is over, crows often forage in large groups of thousands of birds and share nightly roosts with additional groups.

▲ Crows are found throughout the continental United States and southern Canada. The population is estimated at about 30 million.

Common Raven
(*Corvus corax*)

DESCRIPTION

Ravens average 24 inches long, with a 53-inch wingspan, and weigh about 2.6 pounds. Larger than crows, they are distinguishable by a heavier, larger bill; longer and narrower wings; a long, wedge-shaped tail; and a hoarser call. Ravens have glossy black coloring with a shaggy ruff of feathers or **hackles** on the throat. Ravens also soar and glide more than crows do.

The smaller Chihuahuan raven is found in areas around western Texas.

HABITAT AND BEHAVIOR

Ravens are often found in hilly or mountainous forests; near open areas of grassland, deserts, and scrubland; near open seacoasts or riverbanks; and in the tundra. They are found less often in human areas than crows, although they can adapt to both agricultural and urban settings; in some areas, in fact, they have become agricultural pests. Most often living as solitary individuals or nesting pairs, ravens less commonly form foraging groups and roosting flocks, depending on food resources. They roost in large trees, cliff sides, electric towers, and manmade structures.

It is unknown whether pairs mate for life, but from late winter to early spring the female lays 3 to 7 eggs. The pair aggressively defends the eggs and nestlings from raptors, owls, and other predators. The young may leave their parents by 7 weeks of age or remain with parents somewhat longer. Mature at age 3, wild ravens live an average of 13 years, although ravens in captivity live far more than 40 years.

Ravens are primarily scavengers of mammal carcasses and the insects that populate them, approaching cautiously after vultures have torn carcasses open or crows or jays have begun to feed. Omnivorous hunters and gatherers, they will consume the afterbirth of livestock, pick through animal dung and garbage, prey on eggs and nestlings, and store or cache food.

wingspan
4⅜ feet

24 inches long

▲ Common Raven

▲ Ravens are found in the western states north into Canada and Alaska, and east from upper Canada to New England, including Ontario, Quebec, northern Minnesota and Wisconsin; and south into the Appalachians. Birds living in arctic areas do migrate, while most other ravens are year-round residents. As the population recovers, ravens are beginning to return to former areas in eastern and middle states, and the northern plains.

The raven's large, heavy bill gives it a distinctive profile.

213

Black-Billed Magpie
(*Pica hudsonia*)

DESCRIPTION

The black-billed or American magpie averages 19 inches in length, with a 25-inch wingspan, but weighs only 6 ounces. Males are somewhat larger and heavier than females. This magpie has a black head and neck; white shoulders and belly; a large white area on the wings; and iridescent blue or blue-green on the body, wings, and tail. The tail is very long, forming half the length of the bird, and the feet and bill are black. The closely related yellow-billed magpie is found only in California and is distinguished by a yellow bill and a yellow streak around the eye.

Wingspan 25 inches

17¾–23½ inches long

▲ Black-Billed Magpie

HABITAT AND BEHAVIOR

Originally associated with the great bison herds, magpies are now often found in close association with cattle on rangeland, where they perform the same function, helpfully eating ticks and other insects off the animals. Magpies also scavenged the carcasses of bison hunted by Native Americans. Although preferring open areas with scattered shrubs and trees, they can also be found in pine forests, farmland, and more suburban areas. Unless frightened, magpies can be bold or nearly tame with humans.

Although they are mostly permanent residents, some birds may descend to lower elevations in colder months. Nesting territories vary in size depending on food resources. The birds form loose family flocks of 6 to 10 birds. Several hundred birds come

Black-billed magpies build large, canopy-covered, basket-shaped nests high in trees.

together in social roosts in winter, perching and sleeping in individual trees.

Hunting and foraging. Walking on the ground, magpies eat primarily insects but also eggs, small hatchlings, rodents, seeds, berries, fruit, and grain. They cache food in the dirt and cover it with debris. Magpies also scavenge kills from larger predators, and pick through garbage, animal feed, and dung. They will pick at wounds on cattle or the eyes of newborns, kill nestling poultry, and eat eggs.

LIFE CYCLE

Magpies are monogamous, sometimes mating for life. Breeding season runs from March to July. Pairs build an elaborate domed nest high in a tree. The female incubates 6 to 7 eggs, but within 2 months the young leave to join other juveniles. Reaching maturity at ages 1 or 2, they do not live long in the wild but in captivity will live to 20 years. Raptors, owls, crows, ravens, and many small predators prey on magpies.

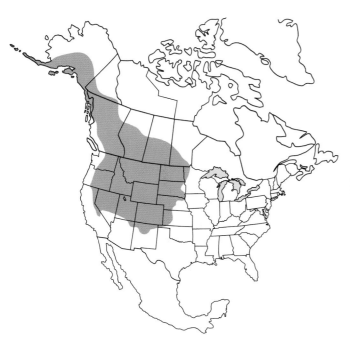

▲ Magpies are found in the interior of the western half of North America, from southern Alaska, through western Canada, south to northern California, Arizona, and New Mexico.

HUMAN INTERACTION

The corvids eat a large variety of insect pests and are important scavengers of carrion, but they can also eat fruit, nuts, and grain crops. They are known to kill young goats, lambs, and calves by pecking at their faces and plucking out their eyes. They also attack the young of poultry, eat eggs, and peck at sores on livestock. Many members of the crow family will dive-bomb humans or other animals in defense of their nest.

LEGALITIES

The Migratory Bird Treaty Act protects crows, ravens, and magpies. It is unlawful to hunt, capture, or harass them, or take their eggs. Ravens are listed as Endangered in some states. State and local laws vary.

Dealing with Corvids
Homes and Yards

- Remove attractants such as garbage, carrion, bird feeders, and pet and animal feed.

- Protect fruit and vegetable crops, poultry, and small animal pens with netting.

- Remove low brush and trees and all old nests to minimize magpie nesting.

- Thin tree branches to reduce roosting habitat.

- If you can't avoid a nesting area on your property, carry an umbrella as protection against aggressive diving attacks.

Livestock Husbandry

- Keep poultry, especially nesting birds and hatchlings, in protective, covered pens and coops.

- Install visual scare devices (such as Mylar tape, balloons, or flags) to provide temporary protection for newborn livestock, poultry, and crops. Use various scare devices in various combinations, and change sites and devices.

- Set up motion-activated water and sound devices. Recorded distress and warning calls will temporarily deter birds.

DAMAGE ID: Crows

PREY ON

Young goats, lambs, calves, the young of poultry, eggs

TIME OF DAY

Birds begin foraging very early in morning, and all attacks occur during day.

METHOD OF KILL

Because corvids are primarily scavengers, they will be attracted to dead or dying animals and birthing. This behavior can be misidentified as predation.

- ▸ Repeated pecking at animal sores and damage to the face, skull, or eyes of live animals, including bloody eye sockets. If the animal was already dead, the eye sockets will not bleed when eaten by scavengers.

- ▸ Only small bones broken, because corvids lack strong enough beaks and talons to make quick kills or tear open skin

- ▸ Shell pieces left after egg eaten, or a jagged hole pecked in the side of the egg, leaving it fairly intact

TRACK

Crow: Classic, 2½ inches long and 1¾ inches wide.

Raven: Classic, 3¾ inches long and 2 inches wide.

Black-billed magpie: Classic, 2 inches long and 1½ inches wide.

Claw marks visible

Crow   Raven   Magpie

GAIT

Crow: Walking stride 5 inches

Raven: Walking stride 20 inches

Black-billed magpie: Walking stride 6 inches

SCAT AND PELLET

Semiliquid, brown, black, and white, but varies with diet

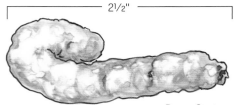

Crow Pellet   Raven Scat   Magpie Pellet

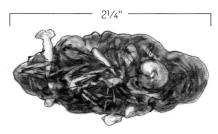

Raven Pellet

CHAPTER 15

Snapping Turtles

Chelydridae

Named for their distinctive method of capturing prey, snapping turtles have a reputation for fierceness in self-defense, since they are unable to retreat into their shell. Members of this family have very large limbs and heads, strong jaws, and long tails.

Typically, snapping turtles move quietly or lie in ambush waiting for prey, and then suddenly and rapidly lunge with an open mouth and extended neck, snapping down on their target with a cutting bite.

The freshwater snappers, the survivors of a family with many extinct members, are found in North and South America and are divided into two genera — *Chelydra* and *Macrochelys*. The only other member of the snapping turtle family, the big-headed turtle, is found in southeast Asia. This turtle lives in mountainous streams but is a nocturnal forager on land and an excellent rock climber rather than a swimmer.

Common Snapping Turtle
(*Chelydra*)

The common snapping turtle is sometimes further divided into two subspecies: the northern snapping turtle (*C. serpentina serpentina*) and the Florida snapping turtle (*C. serpentina osceola*). Two other species of snapping turtles live in Central and South America.

DESCRIPTION

The **carapace**, or upper shell, of the common snapping turtle can variously be green, tan, or brown to black, with rough scales on the top and saw-toothed edges toward the back. The *plastron*, or lower shell, is too narrow and small to allow the turtle to pull its legs, tail, and head into the shell space. The yellowish legs are sturdy and thick. The yellowish tail is almost as long as the shell with saw-toothed ridges along the top. The large, dark-colored head has powerful jaws.

Turtles are measured by the straight carapace length from the front to the back of the shell. The common snapping turtle varies from 8 to 18 inches in length, with weights of 10 to 35 pounds. Males are larger than females.

HABITAT AND BEHAVIOR

Common snapping turtles prefer freshwater wetland environments with vegetation and muddy bottoms such as ponds, lakes, swamps, marshland, or slow-moving rivers and streams. In Florida they can be found in brackish water as well. In areas of colder

Adult carapace 8-18 inches

▲ Common Snapping Turtle

Common snapping turtles can be identified by sawtooth ridges on the upper shell, a very large head and neck, and a long, ridged tail. With age, the ridges become smoother.

water, snappers will bask in the sun but not as often as other turtles. In shallow water, they often lie in ambush with only their snout exposed.

Opportunistic omnivores, snappers consume aquatic vegetation, fish, amphibians, crustaceans, insects, reptiles, waterfowl, small mammals, and carrion. Once the water temperature becomes colder, snappers stop feeding and settle to hibernate in depths of 12 to 39 inches, buried in sand or mud with only their eyes and snout exposed.

Snappers lead solitary lives except during mating, which occurs about 6 weeks before the female lays one clutch of 26 to 55 eggs. In the warmer southern states, egg laying occurs from mid-May to mid-June, and in northern areas about a month later. A female can travel as much as 9 miles in search of sandy areas with little vegetation and good sun exposure, traveling

about 165 yards in a day. She then scrapes out a shallow hole for her eggs, and hatching occurs from late summer to early fall depending on incubation temperature. This is when farmers and homeowners see snappers out of their usual environment.

Females mature at as early as 12 years old in warmer areas or as late as 20 years. They may live over 100 years. Large predators such as alligators and bears will prey on mature snappers, and many smaller predators, birds, snakes, and large fish eat juveniles.

Alligator Snapping Turtle
(*Macrochelys*)

Compared with the common snapper, the alligator snapping turtle has a small range. Some experts divide the *Macrochelys* into 3 separate species: the Apalachicola (*M. apalachicolae*), the Suwannee (*M. suwanniensis*), and the alligator snapper (*M. temmincki*).

DESCRIPTION

One of the heaviest freshwater turtles in the world, the alligator snapper ranges from 13 to 32 inches long and weighs from 8 to 250 pounds, though averaging 30 to 50 pounds. It is easily identifiable by 3 large, pointed ridges along the top of the carapace and spikes or serrations along the edges of the shell. Shell color can be solid brown, black, gray, or olive green. The jaws are powerful, the head large, heavy, and distinctive with the eyes on the sides. The alligator snapper's relatively shorter neck can make it easier to handle than the common snapper, but it is still potentially very dangerous.

HABITAT AND BEHAVIOR

Adult alligator snappers tend to be found submerged in deeper water of lakes, river, canals, and swamps. Young turtles often live in smaller streams or ponds. Both active hunters and scavengers, these snappers eat fish primarily but also a wide range of animals, fish

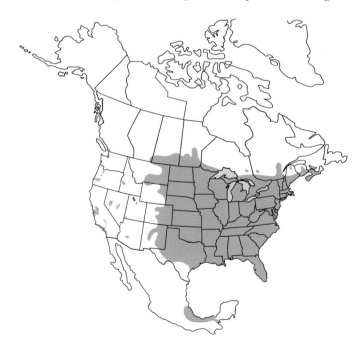

▲ Common snapping turtles were originally confined to an area across the continent to the eastern edge of the Rocky Mountains, northward into southeastern Canada up to Nova Scotia, and south to the Gulf of Mexico. As an introduced, invasive species, they are now found in all the continental states.

carcasses, carrion, and some plant life. During the day they lie in ambush, mouth open, and do more active hunting at night. Inside its mouth, the alligator snapper has a pink **vermiform**, or worm-shaped, appendage on the tip of the tongue, used to lure fish.

Alligator snappers mature at about 12 years of age, surviving up to 45 years and sometimes quite a bit longer. They mate in early to late spring, emerging on open land only to lay 8 to 52 eggs. About 2 months later, the female digs a nest about 160 feet from water. Hatching occurs in fall and is dependent on incubation temperature.

Adult carapace 13–32 inches

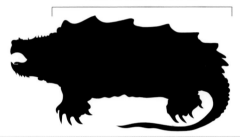

▲ Alligator Snapping Turtle

▲ Alligator snapping turtles are generally found in the river systems of the Mississippi River basin, the Gulf States, and Texas, and as far north as Indiana, Illinois, Kansas, and Missouri.

The alligator snapper has a triangular, pointed head. When its mouth is open, the fleshy vermiform lure is visible.

HUMAN INTERACTIONS

Snapping turtles will opportunistically prey on waterfowl, including ducks and geese, and occasionally other poultry or small animals or pets.

Snapping Turtle Encounters

A snapper's bite can be serious. Depending on the turtle's size, its bite can easily remove a finger or snap a bone. Its claws are also sharp and substantial. When snappers feel threatened they usually hiss before biting, but they prefer to avoid confrontations.

Common snappers should be picked up only with great care since their head and neck are capable of whipping across their back to their hind feet. Do not lift the turtle by its tail, which can injure the spinal vertebrae. The safest method is to use a square or snow shovel to lift the turtle, placing it on a tarp, which can be carried from the sides by two people. Experienced handlers also pick up snappers using two hands placed between each rear leg and the tail. Due to their shorter necks, alligator snappers can be handled by carefully grasping their shells.

LEGALITIES

Contact state conservation or wildlife agencies for legal status for both common and alligator snapping turtles, which are protected in some areas.

Dealing with Snapping Turtles
Fencing

- Bury 1-foot-high mesh fencing 6 to 8 inches into the soil to exclude snappers from areas you wish to protect.

- Remove muddy bottoms of ponds, and make banks vertical.

- Place large rocks to prevent access to birds' potential nesting sites.

DAMAGE ID: Snapping Turtles

PREY ON

Ducks, geese, other poultry, small animals, pets

METHOD OF KILL

▸ Legs of small or large birds mangled or missing

▸ Ducklings or goslings missing

▸ Nests raided for eggs or whole small birds taken

▸ Trails in soft soil showing a tail drag, foot and claw prints, or just claw impressions

SCAT

Undefined, soft, green–black

TIME OF DAY

Nocturnal

TRACK

Front, 5 toes with claws, roughly oval–shaped foot pad. Rear, 4 toes with claws, long pad almost bear-like in shape. Substantial tail drag mark between left and right footprints. Size varies.

GAIT

Walking stride about half of carapace length, rear feet usually understep

Gators and Crocs

Crocodilia

Descendants of the dinosaurs, the large, semiaquatic reptiles in the ancient order *Crocodilia* appeared 250 million years ago. Alligators, crocodiles, caimans, and Indian gharials all belong to this order. The alligator and crocodile have bodies that are armored with bony plates called *osteoderms* or *scutes*.

The Alligatoridae family has only two living members — the American alligator and the smaller Chinese alligator. The American alligator is related to the caimans found in Central and South America. Alligators are considered less dangerous to humans than crocodiles.

The 14 species of "true" crocodiles, or Crocodylidae, are found in tropical areas in Africa, Asia, Australia, and North and South America. The larger species are very dangerous, killing hundreds of people in Asia and Africa each year. The American crocodile is found in the United States only in southern Florida and Puerto Rico.

AMERICAN ALLIGATOR
(*Alligator mississippiensis*)

When European explorers first encountered the American alligator, they confused it with the very real and legendary Old World Nile crocodile. The name *alligator* came from the Spanish *el lagarto*, or "the lizard," which became *alagarto* in Florida. Soon gaining the nickname *gator*, they were hunted into near extinction in the 1950s and '60s. With protection, the population has now very successfully recovered.

DESCRIPTION

The **osteoderms** or **scutes** on the alligator are olive-brown or black with an off-white underside. Young alligators have yellow stripes on their tails. The snout is broader than that of the crocodile, and the large fourth tooth in the lower jaw is not visible when the mouth is closed, unlike the crocodile. Although an alligator is mature when it reaches 6 to 7 feet in length, females will grow to 9 to 10 feet long and males to 13 to 14 feet long, weighing 800 pounds. Larger animals are frequently seen. Females take 10 to 15 years to mature, and males 8 to 12 years. Life span is uncertain but ranges from 35 to 50 years in the wild and as long as 80 years in captivity.

Alligators swim, propelled by their long, strong tail, at speeds as high as 10 mph. They can dive for long periods. On land they either crawl or **sprawl** and can rise up on their legs for a high walk, with speeds of 7 to 9 mph for short distances. Alligators will crawl to new bodies of water if necessary. They can stand and step forward in a lunge. They are also strong climbers.

Alligators have a wide range of vision and are excellent at seeing and sensing movement. They also have a good sense of smell. Their jaws are extremely powerful in biting and gripping but comparatively weak in opening against restraint.

Most active when the temperature is between 82 and 92°F, alligators will bask or seek warmer water to keep themselves warm.

HABITAT AND BEHAVIOR

Alligators live in fresh or brackish water, including ponds, lakes, rivers, wetlands, and manmade canals. They dig holes in wetlands, which provide habitat and water for fish and wildlife in dry conditions.

Alligators stop feeding when the temperature drops below 70°F and become dormant below 55°F. During this time, they retreat to dens or long tunnels near open water, keeping their nostrils just above water to breathe, even if their upper body becomes frozen in ice.

Adult 6–14 feet

▲ American Alligator

▲ Native to the southeastern United States, alligators are currently found in all of Florida and Louisiana; southern areas of Alabama, Georgia, and Mississippi; eastern Texas; coastal areas of North and South Carolina; and up into the southeast corner of Oklahoma and southern areas of Arkansas. Over 1 million alligators live in both Florida and Louisiana, their most important range.

Male alligators are solitary and territorial. Large females will also defend their territory and offspring, while young alligators usually share areas. Alligators begin to mate in April through May or June, with the female eventually building a nest and laying 32 to 46 eggs. Hatching is dependent on incubation temperature, but usually occurs in late August or early September. Meanwhile the mother protects her nest, hissing and charging at threats, and she will continue to watch over her young hatchlings for a year or more.

Alligators hunt primarily at night, submerged in water waiting for prey. They consume smaller food in one bite and will attempt to tear off pieces of larger prey by violently spinning or shaking it. Large prey is also cached underwater until it partially rots and can be more easily eaten. Young alligators eat crustaceans, fish, snails, worms, and similar small animals. Older alligators consume larger fish, turtles, snakes, birds, and both small and large mammals such as deer.

Alligators are a very important predator of the very destructive nutria and muskrat. Mature alligators are able to capture very large predators such as bears and Florida panthers. Young alligators themselves are prey for birds, snakes, fish, small predators such as raccoons, otters, and larger alligators. Mature alligators are apex predators that are threatened only by each other in fights or by human hunting.

HUMAN INTERACTIONS

Attacks on people and pets are increasing as humans move into alligator habitat, especially next to water. Any poultry, waterfowl, or livestock that is allowed to freely roam in alligator areas is very vulnerable.

LEGAL STATUS

Federal protection is due to the similarity of appearance with endangered American crocodile. Licenses or permits are required to trap or kill alligators. Contact state alligator nuisance control programs or wildlife agencies for removal.

Feeding alligators is often illegal because it greatly increases the danger to humans.

AMERICAN CROCODILE
(Crocodylus acutus)

Found more commonly in the Caribbean, Central America, and South America, American crocodiles living in southern Florida are at the extreme northern part of their range. Since the American crocodile's listing as a federally Endangered species in 1975, the population has increased from a few hundred to about 2,000.

DESCRIPTION

American crocodiles have **osteoderms** or **scutes** on their back and a long, strong tail. Lighter colored than alligators, crocodiles are grayish green with yellowish-white undersides. Young crocodiles are lighter tan with dark stripes. They are distinguished from alligators by their narrow head and tapered snout, and the fourth lower tooth that is exposed when the mouth is shut. The ears, eyes, and nostrils are located on the top of the head, so that they are exposed while the rest of the crocodile is submerged. The front legs have 5 toes on each foot and the rears have 4 toes. Mature males can reach 14 to 15 feet long with weights of 800 pounds or more, while females are generally 8 to 12 feet long. Average life expectancy is 60 to 70 years, although some individuals reach 100 years or so.

HABITAT AND BEHAVIOR

Coastal brackish and saltwater estuaries, swamps, creeks, and ponds are their primary habitat, but crocodiles have recently moved through freshwater canals into other areas. Shy and reclusive in nature, crocodiles construct burrows for resting and

The crocodile's distinctive features include a large, lower tooth that extends out of the mouth and over the upper jaw, as well as a narrow, V-shaped head and snout.

protection in cold weather. Provided with sufficient food, crocodiles are solitary and do not leave their own area except during mating season. In late April and early May, females build their soil nests above water, lay 20 to 60 eggs, and return near hatching time in late July or early August. After assisting the hatchlings for a day or so, the female leaves the area.

Crocodiles bask in the sun and are sometimes seen with their mouths open. This gaping behavior is related to regulating their body temperature and is not aggressive. During dry seasons, crocodiles will bury themselves in mud and become lethargic.

Adult 8–15 feet

▲ American Crocodile

Crocodiles will submerge themselves in water during cooler weather but are unable to survive in temperatures below 45°F.

Primarily feeding at night, young crocodiles eat crustaceans, small fish, and amphibians; while larger adult crocodiles feed primarily on large fish, and less commonly on turtles, snakes, birds, small mammals, and the occasional deer. At any time, crocodiles may lurk below the surface waiting for animals that approach the water's edge. Although large fish, raccoons, and larger predators frequently prey on juvenile crocodiles, adult crocodiles are apex predators.

HUMAN INTERACTION

Conflicts between crocodiles and humans are rare, partially due to their low numbers. The first human attack in Florida occurred in 2014. Crocodiles will take dogs, goats, pigs, or cattle if available. Crocodiles can be more aggressive toward humans than alligators are.

LEGALITIES

Crocodiles are a federally protected Endangered species.

Alligator or Crocodile Encounters

Pay close attention for the presence of alligators or crocodiles when you are near water at dusk, dawn, or during the night and in the warmer months. Alligators living in waterways or ponds near housing or golf courses can become dangerously habituated to humans, even during daylight.

Attacks occur most often when a person is attempting to pick up or capture an alligator, followed by swimming, fishing, and retrieving golf balls. An average of 5 human attacks occur yearly, with at least 19 deaths since 2000. Dogs and other small animals are attacked far more than humans. Seek treatment for any bite to prevent infection from the large amounts of bacteria in an alligator's or crocodile's mouth.

▲ In the United States, crocodiles are found only in tropical southeast Florida.

Caimans (*Caiman crocodilus*)

Spectacled caimans, probably former pets, were first found in southern Florida in 1960. They are primarily found in freshwater lakes and canals. Looking more like crocodiles, caimans are olive brown, with more pointed snouts than alligators. Invasive caimans have not exceeded 6 feet in length. Caimans are able to reproduce in the very southern areas of Florida and do not survive winters north of central Florida. Caimans are aggressive if confronted or threatened.

Caiman breeding populations are now found in Broward and Dade counties, with individuals sighted elsewhere.

Recreation

Practice safety measures when hiking, kayaking, fishing, hunting, or camping. Inquire about current alligator or crocodile activity before recreation in known habitats.

CAMPING

- Do not camp next to water. If possible, campsites should be 6 feet above the waterline and 150 to 160 feet from the water's edge. Avoid areas with wallows, mudslides, and nests, as alligators are very territorial.

- Do not wash dishes or dispose of garbage in the water. Dispose of all garbage appropriately in trash containers where possible.

- Do not take your dog camping in alligator or crocodile areas.

FISHING

If you encounter an alligator or crocodile, or it is attracted to your lure, stop and move away. Do not release fish where alligators or crocodiles are present, to prevent association with humans. Dispose of fish scraps or waste appropriately.

HUNTING

Do not clean game meat or discard entrails in or near water or near your campsite.

KAYAKING

Be extremely cautious of shallow, narrow areas 5 to 10 feet wide. Do not cut around bends too closely or enter shallow areas where alligators like to hide and wait for prey. If an alligator or crocodile is observed, try to keep a distance of 100 feet. Use great caution when entering and leaving the water.

Dealing with Alligators and Crocodiles

The following advice refers to alligators. Although quite rare, crocodiles behave very similarly and the same precautions are necessary in their areas.

Homes and Yards

- Do not feed alligators. Feeding alligators encourages them to associate people with food.

- Set up wooden or concrete waterside bunkers, at least 3 feet tall, to discourage alligators. Fences taller than 5 feet are the best protection against alligators in yards or areas with children, pets, poultry, or livestock. Alligators can climb mesh or chain link fences up to that height. Fencing should have a minimum of 4-inch-wide openings and an outward, angled overhang.

- Do not allow small children and pets to play unsupervised in alligator areas, especially near the water's edge.

- Keep your dog on a leash. Alligators often wait offshore for prey to approach the water, then grab its head and pull it underwater.

- Do not wade, dangle your feet, swim, or allow pets to swim in known or posted alligator areas. If you fall into the water, do not splash or make noise. Wade or swim to shore quietly and quickly.

- Do not swim with your dog, as dogs are definitely seen as prey.

- Do not approach an alligator, its nest, or baby or juvenile alligators.

- Do not harass or attempt to capture alligators. Contact authorities for removal of a nuisance alligator.

- Do not make pets of small alligators, which will become habituated to humans.

Alligator Attacks

- If an alligator hisses, leave immediately.

- If an alligator charges, run fast in a straight line away from it and the water, and not a zigzag pattern, which is mistaken advice.

- Most attacks by smaller alligators consist of a single bite and release. Alligators of 8 feet in length or more usually make serious attacks.

- An attacking alligator will attempt to drag you into the water. Fight back at its eyes and head. Alligators usually grab an appendage and attempt to twist it off.

- If you observe an attack on a person, continually hit the alligator on the head with a heavy stick or another object.

DAMAGE ID: American Alligator and American Crocodile

PREY ON

Poultry, pets, goats, pigs, cattle

TIME OF DAY

Primarily night

METHOD OF KILL

- Lie submerged in water; will also lie in shallow water
- Small animals entirely consumed
- Prey taken underwater to drown
- On larger animals, portions of carcass torn violently off
- Larger prey cached underwater

TRACKS

Front, 5 long toes with one facing to the rear. Rear, 4 long toes and long heel pad. Claw marks visible. Walking tracks may show the drag of the tail and the prints of all four feet; or the tail may obscure the footprints.

GAIT

On land, either crawl or rise on hind legs to walk short distances. Crawl is an alternate diagonal pattern, with front left and rear right moving together. Crawl may be slow with stomach and tail also on ground. At faster speeds body lifts off ground and tail moves quickly from side to side.

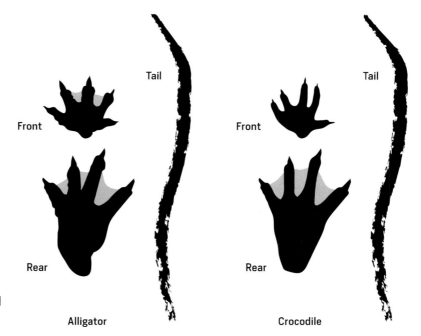

Alligator

Crocodile

SCAT

White in color, scat primarily consists of undigested calcium from bones and shells.

Alligator

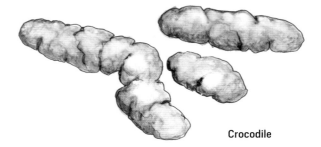

Crocodile

229

Snakes

Colubridae

Carnivorous reptiles without legs, snakes are *ectothermic* (commonly called "cold-blooded") and covered with smooth, dry scales. Their low, flexible jaw and the joints in their skull allow them to swallow prey larger than their own body. Specialized belly scales help them grip and travel.

There are 160 species of snakes in North America. The most common families are:

- **Colubridae**, including the rat snakes, king snakes, water snakes, and garter snakes, among others

- **Viperidae**, including the copperheads, cottonmouths, and rattlesnakes

- **Elapidae**, the coral snakes

Most snakes are nonvenomous and harmless to humans, performing valuable environmental service as predators of rodents and other pests.

Snakes have a directional sense of smell and taste. The forked tongue collects particles from the air, ground, and water and transfers them to the **vomeronasal** organ in the mouth, which is an auxiliary smell sensor. They can also sense vibration through their body's contact with the ground. Their eyesight is variable but not especially sharp.

Found in the Northern Hemisphere, snakes in the large Colubridae family lack hollow fangs, have large symmetrical head plates, and are constrictors, primarily feeding on rodents and birds. Most pose no threat to humans.

Rat Snake and King Snake

DESCRIPTION

Among the most common predators of eggs, poultry, and rabbits, the various rat snakes (*Pantherophis*) and king snakes (*Lampropeltis*) are nonvenomous. They can vibrate their tail like a rattlesnake when alarmed.

Reaching 3 to 6 feet in length, the common king snake can be nearly completely black or display vibrant colors and patterns — leading to confusion with venomous coral snakes. Found near wetlands, king snakes also inhabit fields, wooded areas, and abandoned buildings. In addition to eggs, lizards, rodents and birds, they eat many venomous snakes.

The common rat snake is also medium sized. Although the subspecies are seen in various colors, black rat snakes are the most widely distributed. Excellent swimmers and climbers of trees and even brick walls, the various subspecies of rat snakes have adapted to homes from the Everglades to the mountains. Rat snakes are generally welcome around barns since they primarily eat rodents, chipmunks, and squirrels, but also eggs and nestlings.

HABITAT AND BEHAVIOR

Depending on the species, snakes are found in many different habitats. They often hide or lie quietly if they are observed, hissing or striking only if cornered. Snakes will bask in the sun to gain warmth, but seek shelter to avoid overheating during the hottest times. While some snakes are more active during the night, others, such as king snakes, prefer the morning and the day. Snakes usually move slowly but are capable of faster efforts over short distances to escape a threat or to chase prey.

In colder regions, snakes are awake but dormant and inactive during winter, either alone or in a group den known as a **hibernaculum**. Often used for many years, these dens can be found in old animal burrows, under stumps and logs, in rocky areas, and in

SPECKLED KING SNAKE

PRAIRIE KING SNAKE

BLACK RAT SNAKE

lumber or trash piles. Mating occurs after the snakes emerge in spring, one of their most active times. Depending on the species, most hatch from eggs although some young are born live. Snakes generally reach maturity in 2 or 3 years and may live for 20 years. They will shed or molt their skin periodically as they grow.

Snakes are completely carnivorous, consuming many pest animals such as mice, rats, voles, snail, and slugs in addition to insects, eggs, nestling birds, frogs, lizards, and fish — swallowing their prey whole. After they eat, snakes become dormant to allow for digestion. If disturbed, a snake may regurgitate its meal in order to escape from a threat.

Snakes are the prey of many small predatory mammals and birds, including chickens.

HUMAN INTERACTION

The best advice is always to leave a snake alone. In potential snake areas, do not step or place your hands into areas you cannot see. Wear high-top boots and long pants when hiking or walking in snake areas. Open or lift items on the ground away from you. Use a stick to disturb areas to warn snakes away. Most snakes that are killed are harmless and nonvenomous. Not all venomous snakes give a warning, especially copperheads.

Venomous Snakes

Twenty species of venomous snakes — rattlesnake, cottonmouth or water moccasin, copperhead, and coral — are found in the United States, some in every state except Alaska. If these snakes live in your area, learn to identify them.

Treatment for a bite from a venomous snake requires medical attention. Although identification is important, do not stop to catch the snake since antivenin can treat bites from rattlesnakes, copperheads, and cottonmouths. While several thousand people receive bites every year, deaths are very rare.

Bites from nonvenomous snakes may cause damage or infection but are generally harmless.

TIMBER RATTLESNAKE

COTTONMOUTH

The release of former pet snakes is responsible for the growing problem of very large pythons, constrictors, and anacondas in Florida. The National Park Service estimates that 100,000 Burmese pythons, which routinely grow to 20 feet long and 250 pounds, now live in the Everglades. Attacks on pets, small animals, and poultry are increasing. The US Fish and Wildlife Service is conducting ongoing research into containing the expanding population and range.

Homemade Snake Trap

To relocate a snake, you can make a trap from a 2- to 3-foot section of 4-inch PVC pipe, capped at one end. Anchor the trap to prevent rolling. Entice the snake with a heat or cold source (depending on the temperature), wrapped in a clean rag. Sprinkle flour in front of the trap to reveal a track if a snake has entered. Cover the opening with duct tape and remove the trap.

LEGALITIES

Depending on the state, many snakes are protected by law unless there is an immediate threat. Some snake species are protected or Endangered species.

RANGE

King snakes have the largest range of any North American snake species, found throughout most of the United States into southeastern Canada.

Rat snakes are found from the southeastern states, Texas, and Oklahoma, to southern New England through to Michigan and Wisconsin. There is a small, threatened population of black rat snakes in southern Ontario.

Dealing with Snakes
Homes and Yards

- Eradicate mice and other rodents.

- Remove potential hiding spots of snakes and rodents, such as tall vegetation and wood, rock, and trash piles.

- Secure compost bins instead of leaving open compost piles.

- Keep in mind that damp mulch and large landscape rocks are more appealing to snakes than smaller river rocks or pebbles.

- Keep grass short and trim shrubs at least 6 inches up from the ground.

- Do not chase or attempt to capture a snake because you may force it into hiding. If a snake has entered a building, open a door so that it leaves on its own or with gentle herding.

- Use a long pole and container to move or trap snakes.

- Be aware that sulfur, lime, mothballs or naphthalene, hot pepper, king snake musk, and many other products do *not* repel snakes.

Livestock Husbandry

Snakes usually enter structures at ground level, through holes or cracks and under ill-fitting doors.

- Place coops and hutches aboveground to limit snakes' access.

- Cover all potential entry and ventilation locations with ¼-inch hardware cloth, patch or fill holes, and add tight-fitting weatherstripping to doors.

- Install a 12-inch coop apron and solid flooring to prevent access through rodent holes.

- Remove tree branches or vines that snakes can climb.

- Free-range poultry is more difficult to protect, although nesting boxes can be housed in predator-proof tractors or coops.

- Protecting waterfowl eggs and nestlings is more difficult if they are allowed free access to bodies of water, rather than secure pens.

Fencing

- Enhance existing fence with a 24- to 30-inch-tall hardware cloth on the outside bottom, burying another 2 to 4 inches into the soil.

- Construct snake-deterring fencing of 24- to 30-inch-tall hardware cloth slanting 30 degrees outward, with an additional 3- to 6-inch underground deep barrier, and posts placed inside the fence to prevent climbing.

- Make sure that gates are tight fitting with similar hardware cloth reinforcement.

- Keep vegetation mowed near the fence and fill in any animal holes.

DAMAGE ID: Snakes

PREY ON

Eggs, chicks, nestlings

METHOD OF KILL

▸ Egg, chick, or nestling swallowed whole; a shiny slime left on a dead bird that is too big to swallow.

▸ Snake present after kill, dormant or unable to fit out of opening

TRACK

Track may be fairly straight, undulating, or sidewinding, depending on the species.

TIME OF DAY

Day

SCAT

Snake scat can be black or brown, and white nitrogenous material may be present. You may also find snakeskins.

Predators in Hawaii

Even though Hawaii lacks most of the predators found in North America, there are still a few animals and birds that can attack poultry or livestock. Feral cats, dogs, and pigs pose a threat, as do the rats that made their way to the islands. A variety of predatory or troublesome birds include hawks, eagles, falcons, jays, crows, magpies, and ravens. Poultry raisers report that feral cats, mongooses, and predatory birds are their major problems. Free-range poultry is at greatest risk. Secure housings and pens with covers are required.

'Io or Hawaiian Hawk (*Buteo solitarius*)

The only hawk native to Hawaii, the 'io is currently breeding only on the Big Island but is occasionally seen on Kauai, Maui, and Oahu. Its population is estimated at around 3,000 birds. The bulky body is 16 to 18 inches in length, with broad wings, and weighs about 21 ounces. The bird has a dark head, a dark or light breast and underwings, and yellowish feet and legs. The female is larger than the male. Mating for life, the female lays 1 egg yearly.

The 'io feeds on young or small chickens in addition to rodents, insects, and other small birds. It usually hunts from a perch but may also dive on prey.

Pueo or Hawaiian Short-Eared Owl (*Asio flammeus sandwichensis*)

This brown owl, 14 to 17 inches long, is found throughout the islands, but its actual population is unknown. The pueo nests on the ground, laying 3 to 6 eggs over several months. The parents raise the different-aged nestlings together. They are very vulnerable to feral cats and mongooses.

Active during the day, the owl hunts small mammals such as rats, mice, and mongooses, but chickens are a favored prey.

Small Asian Mongoose (*Herpestes javanicus*)

Sugarcane farmers introduced 72 mongooses in 1883, under the misguided idea that they would hunt rats on the Big Island. Unfortunately, the rats are nocturnal and the mongoose is active in the day. Today this invasive species is found on Molokai, Maui, Oahu, and Hawaii, and has possibly spread to Kauai. It has no natural predators on the islands. The mongoose is extremely destructive of native birds and turtle eggs and also eats small mammals, reptiles, insects, fruit, and plants. It kills chickens and nestlings and steals eggs.

Mongooses are brown, 8 to 25 inches in length, with a long body, short legs, and a very long tail. They sleep in dens at night and raise 2 or 3 litters of 2 to 5 pups each year. Mongoose-proof fencing is required as they are capable of going through very small openings.

Polynesian Rat (*Rattus exulans*)

Polynesian rats inhabit nearly every Pacific island within 30 degrees of the equator, including the Hawaiian Islands. They are not present on the mainland in North America.

Smaller than the Norway and black rat, they have a slender body, a sharply pointed snout, large ears, and small feet. Both the reddish brown body and the scaly ringed tail are 4.5 to 6 inches long. Typically found in agricultural land, they will consume sugarcane, pineapple, root crops, coconut, corn, rice, fleshy fruits, nestlings, and eggs. They are nocturnal and able to climb trees.

Part III

Prevention and Protection

A predator attack causes a range of emotions, from frustration to great anger. There can be real grief for animals' lives that were your responsibility. There can be a threat to the economic security of your entire farm and all who live there. Most of all, there is an immediate need to protect your other animals and possibly your family.

Much of this stress can be reduced if you plan and practice good prevention techniques rather than just respond to situations as they occur. Lethal responses to predation are generally not effective unless they are highly selective and targeted. Lethal methods — shooting, hunting with dogs, trapping, snaring, aerial hunting, baited cyanide injectors, and livestock protection collars — are all regulated, restricted, or prohibited differently in states, provinces, and local areas. Check with your local department of natural resources and state agriculture authorities, as well as local government, to determine your specific situation.

Nonlethal Prevention Methods and Strategies

Nonlethal approaches to predator control include secure fencing and housing, good husbandry and management, and disruptive and aversive stimuli.

The most successful and effective predator protection programs combine various techniques in a flexible plan suited to different times of year and grazing areas.

Choose the elements that best suit your predator threats, the changing predator pressures during the year, your location, your terrain, and your stock. Predators can become habituated to some techniques, necessitating moving or changing your methods.

What Works

Digging deeper into the US Department of Agriculture (USDA) predator data for cattle and sheep, we can learn a great deal about the current favored nonlethal control methods, where they are working, and what needs to be done. To date, sheep owners have developed and adopted most of the nonlethal control methods, and results are positive. More than half of sheep operations now use one or more nonlethal methods of predator control. The use of these methods has doubled since 2004, and both sheep and lamb predator loss is now less than in any study year in the past 20 years.

The specific strategies that have more than doubled in use include fencing, livestock guardian dogs (LGDs), donkeys, lamb shedding, active herding, night penning, fright tactics, bedding changes, and frequent checks. The use of guardian llamas has decreased. Owners of fewer than 25 sheep use fencing, LGDs, night penning, and lambing sheds in that order. Operations with more than 1,000 sheep, often on range or in large grazing areas, use LGDs, more frequent checking, culling older animals, changing bedding, and fencing, in that order.

Nonlethal predator control methods on cattle operations have been far less utilized, although that practice is also increasing. In 2000, only 5.4 percent of cattle raisers used at least one method; by 2010, 12.4 percent had adopted some techniques. The use of guardian animals, fencing, and carcass removal nearly tripled. The most commonly used strategies included guardian animals, frequent checking, fencing, and culling.

Work remains to be done on increased and improved use of nonlethal techniques. To that goal, specialists from the USDA National Wildlife Research Center's Predator Research Facility are engaged in long-term research projects to place,

Nonlethal Practices

The following techniques not only protect your animals but also fit the guidelines of the Wildlife Friendly Network (see page 270).

- Time pasture use with periods of low predator pressure.

- Time calving and lambing to avoid periods of high predator pressure.

- Use secure birthing areas, such as fenced pastures or paddocks and birthing sheds.

- Use LGDs, guardian llamas, or donkeys with appropriate care and welfare.

- Fence night pens, feeding areas, and bee yards.

- Secure all grain or other nonforage feedstuff.

- Mix smaller and larger animals together.

- Implement electric fencing, fladry, and radio-activated guard boxes (RAG).

- Practice appropriate harassment techniques.

- Closely monitor stock, with herders, range riders, and other frequent human appearances.

- Carefully evaluate and respond to any depredations.

- Remove carcasses.

- Record conflicts or depredations.

- Limit the use of wildlife exclusion fencing except during times of vulnerability, high predation, or for small stock or poultry.

- Use lethal control only with prior authorization, except in the case of an active attack (but not in the case of scavenging by predators).

- Remember that hunting of predators or other recognized key species is prohibited.

Sheep Production	
NONLETHAL CONTROL METHODS	PERCENTAGE OF PRODUCERS USING METHOD
Fencing	31.8
LGDs	23.5
Lambing sheds	20.0
Night penning	19.5
Culling older sheep	9.6
Frequent checks	9.5
Guardian donkeys	8.2
Remove carrion	6.6
Sheepherding	6.4
Changing bedding	6.3
Guardian llamas	5.4
Other techniques	3.9
Change breeding season	2.9
Fright tactics	1.8

Sheep and Lamb Predator and Nonpredator Death Loss in the United States, 2015 (USDA, September 2015).

Cattle Production	
NONLETHAL CONTROL METHODS	PERCENTAGE OF PRODUCERS USING METHOD
Guard animals	4.1
Fencing	3.8
Frequent checks	3.7
Culling	3.0
Carcass removal	2.5
Night penning	0.8
Herding	0.6
Fright tactics	0.3

Cattle and Calves Predator Death Loss in the United States, 2010. (USDA, February 2012)

support, and evaluate the use of LGDs with sheep and cattle producers in the western states. In Texas, experts from Texas A&M University have also instituted an expanded research project center around the use of LGDs with goat producers. As with the USDA researchers, the goals are to place LGDs with ranchers who have not used them previously on their large pastures and to provide training and support for the users. Elsewhere research continues on other methods of nonlethal protection.

Fencing

Fencing is always the first line of defense in predator protection. Keeping stock in is usually easier than keeping predators out, and fences are often designed for that purpose only, not for protection.

Fencing is often divided into two types — **exclusion** and **drift**. Exclusion fencing is designed to keep specific wildlife out. Drift fencing is usually intended to keep specific animals in but often does not provide good predator protection, especially such

choices as board and rail, widely spaced barbed or electric wire, and nonelectrified wire.

Fences that do provide predator exclusion include wire mesh with narrow spaces, multistrand electric fencing, or a combination of these materials. Poultry and small animal fencing in particular needs to be very tight and often utilizes heavy mesh wire screening or hardware cloth, underground aprons, and overhead protection of runs. Properly selected and designed temporary fencing, such as electric tape, wire, rope, or electrified net, can also provide good predator exclusion.

The higher cost of such fencing is justifiable when you are experiencing high losses, large or dangerous predators, or loss of valuable animals. Good predator exclusion fencing will also increase the success of your other prevention techniques, including livestock guardian animals.

Customizing Your Fencing Plan

Some owners choose to create a zone of greater safety to house animals at night or during times of greater vulnerability. Your fencing plan can include choices such as exclusion fencing for boundary areas and drift fencing for interior areas. Others choose to create very secure nighttime shelters and paddocks. The same guidelines are relevant to owners who wish to protect their yard and areas around their home site.

▲ Consolidating your beehives makes the use of bear-resistant fencing more practical. A well-protected and maintained bee yard provides nearly complete security

Wire Mesh Fencing

Why choose it. Wire mesh or woven wire provides several advantages. It serves as a physical and visual barrier and remains effective in power outages.

Tips. The size of the mesh itself and the height are important for specific predator exclusion and stock or dog inclusion. Mesh fencing is available with tighter or smaller bottom mesh for increased predator proofing. Heights of 5.5 to 6 feet tall are highly recommended, but additional electric wires placed 1 foot above a 4- or 5-foot-tall mesh fence will provide good predator exclusion. Adding an electric trip or scare wire on the outside will also prevent digging. If you use LGDs, you can place a scare wire on the inside to prevent digging by a dog as well.

Install mesh fencing tight to the ground and on the inside of the fence posts. **Gauge** refers to the wire size — the larger the number, the smaller and weaker the wire. The different knots used to link the mesh have different characteristics as well.

Best choice. Mesh fencing can be poly-coated or galvanized for longer life. Although more expensive, high-tensile mesh fencing has a longer life span, less sag, and can be strung farther between posts. Galvanized, high-tensile wire is recommended for predator control.

Fencing Out Specific Predators

Coyotes, wolves, bobcats, lynx, and dogs can jump fences shorter than 6 feet tall. Mountain lions can present a greater challenge since they can jump a 10-foot fence. Coyotes can pass through openings as small as 4.5 inches and can often crawl or dig under the bottom wire or under a fence.

Raccoons and some other small predators are excluded by 2- or 3-inch-wide openings, but they can reach through 1-inch openings such as those found in chicken wire.

Bears are often more discouraged by electric fencing than by physical fences unless those are quite substantial, such as livestock panels.

Electric Fencing

Why choose it. Electric fencing is often less expensive than mesh fencing and easier to install on rough ground. On the other hand, it is less visible to animals, requires regular inspection or maintenance, and is not as effective if the power goes out or the fence is shorted. Maintenance is a continual issue.

Tips. The effectiveness of the electric fence is dependent on the voltage, the energy of the pulse, and the degree of contact. The electric fence charger or energizer converts electric power into high-voltage pulses. Energizers can be powered by solar cells, batteries, or the domestic power system. Most chargers are the low-impedance, solid-state type that do very well in moist soil and will shock through green vegetation. Low-impedance chargers work well with large stock. Wide-impedance energizers

deliver higher voltages, especially in situations of dry soil, and work better as both predator deterrence and containment for animals such as goats or poultry. The older high-impedance, or "weed-burner," chargers are more likely to cause fires in very dry conditions and are less effective with animals such as goats, sheep, and poultry.

An electric fence is essentially an open or incomplete electric circuit until it is grounded by an animal touching the fence, which allows the pulsing electricity to travel through its body into the soil to the nearest grounding rod and then back to the energy source. Snow can prevent animals from making contact with ground, and frozen soil doesn't conduct electricity well. Where dry soil, frozen ground, rocks, or snowpack exists, more intense shocks are delivered from "hot/ground" systems that use alternating wires — a charged wire and an uncharged wire are connected to the ground terminal. This grounded wire is a better conductor than soil, delivering a stronger shock. An animal needs only to touch both the charged and the grounded wires to complete the circuit.

Electric fence systems require grounding and lightning protection. All electric fences should be clearly marked to reduce accidental shocks.

Best choice. To discourage predators, an electric fence needs at least 5 wires; increasing the number of wires and the height will create more substantial predator protection. Spacing the wires closer together near the bottom portion of the fence will also help exclude small predators and coyotes. A 6-foot-high, 9-wire fence is an effective predator exclusion fence, with 13 wires providing near-absolute protection. Besides wire, electric fences can be constructed with electric braid, rope, or tape — all of which are more visible than wire to animals. Electrifying barbed wire is considered too dangerous because the barbs can snag people or animals, preventing them from escaping the shocks.

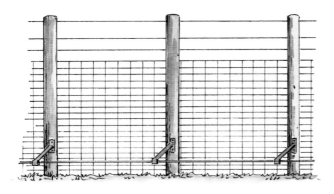

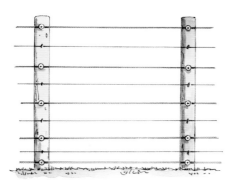

▲ Good predator exclusion uses either tightly woven mesh with the addition of electric wires above and a trip wire outside the fence, or closely spaced, multiple-wire electric fencing.

Puma-Proof Fencing

Since mountain lions (also called pumas) can jump 10 feet high, fences need to be at least 12 feet tall. Heavy woven wire or chain link are more difficult for mountain lions to climb. A wire mesh or electric overhang at the top of the fence is also important. Clear all overhanging tree branches.

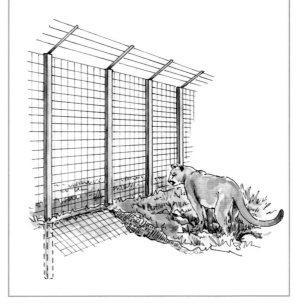

Options for Upgrading an Existing Fence

- Repair all damaged mesh.

- Add height with wires, preferably electric, to discourage predators and climbing or jumping dogs.

- Add an exterior scare or trip wire 6 inches off the ground.

- An interior scare wire will discourage dogs from digging under a fence.

- Add 2 or 3 electric wires outside of the fence to discourage predators.

- Add mesh fencing to an existing board fence.

- Reinforce a board or pipe gate with wire mesh or replace with tightly spaced rail gates.

- Make sure gates fit tightly in their space or create space fillers.

- Add additional water gap or terrain dip protection.

- Predator–proof existing gates by filling in the spaces with livestock mesh, reducing the space under the gate, and adding an overhang.

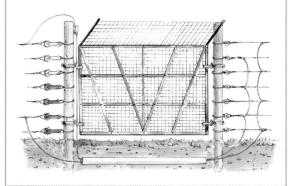

Temporary Fencing

Why choose it. Temporary fencing is most often used to keep stock in a designated grazing area or to create a more protected area for night penning or birthing. It is also useful to contain free-range poultry and to protect beehives and gardens.

Tips. Temporary fencing can be high-tensile wire, poly wire or tape, or poly mesh. Fencing comes in different heights and spacing for specific stock, poultry, or predator applications. Temporary chick fencing is designed to be no-shock. Sagging is reduced with the use of additional vertical struts or posts.

Best choice. Electric netting is the most predator-proof fencing for temporary situations. It will contain sheep, goats, poultry, or dogs and protect against bears, coyotes, bobcats, roaming dogs, raccoons, and other small predators.

It is more reliable and faster to install and move, but it is also more expensive than conventional electric fencing in temporary situations.

▲ Electric netting is a fast, portable, and versatile alternative for predator protection.

Secure Housing for Poultry and Small Animals

The idea of free-range poultry roaming your yard or farmstead makes a bucolic image; free-range birds, however, are at the greatest risk of predation by roaming dogs and other predators. Even in urban and suburban areas, poultry are very vulnerable to predators. If your birds are more pets than livestock, you must seriously consider how to protect them. Whether you are keeping a few birds or raising larger numbers for meat or egg production, consider one of the following general methods of housing and husbandry.

Coop and Penned Poultry

The best way to protect your poultry from predators is to secure them day and night in a predator-proof coop and pen located close to human dwellings, with guardian dogs that will sound an alarm if there is an incursion. Roosters, geese, or guinea fowl can also provide alarm services.

SETUP

Raise the coop off the ground, if possible. Do not plant landscape plants next to the coop or the pen. A raised coop and a cleared area reduce the ability of predators to reach the structure or to hide next to it. This also allows your cats or dogs to patrol under and around the coop to deal with rats, snakes, and other small predators.

The coop needs solid, rat-proof flooring. Have lights you can turn on if you do not leave them on all night. Use only predator-proof latches or lock them closed with a carabiner or something similar, as raccoons can operate most simple closing systems. Owls and hawks will fly into open coop doors, and owls will walk into them as well. If you are not home to close the coop door at dusk, automatic door openers are available using light sensors or timers to shut the door at dusk and open it after dawn.

Chain link, typical farm mesh fencing, and chicken wire are not predator-proof protection for coops or pens. Small predators such as raccoons, weasels, opossums, or rats can easily reach through these materials. The ideal material is 19-gauge hardware cloth or poultry fencing with ¼-inch holes (or no bigger than ½ inch). Use hardware cloth over all windows, doors, gates, open spaces under roof rafters, or methods of ventilation.

To prevent predators from digging under the fencing, you can lay a cement floor, hardware cloth, or closely space livestock panels over the entire surface of the pen and cover with dirt, sand, or deep litter. Run the hardware cloth up the wall or fence of the enclosure. If this is not practical, bury hardware cloth in a trench at least 18 inches deep around pen. Cover the entire top of the pen as well.

▲ The well-designed backyard coop and pen should be tightly enclosed with hardware cloth, with a secure bottom and top.

Free-Range Poultry during the Day

A compromise to a secure coop and pen is to allow birds out to forage during the day and then to secure them at night. Train young birds to return to their coop each evening by keeping them inside their coop exclusively for the first few weeks and by enticing them to return with special feeding right before dusk. They will then consider it their home. Don't feed outside the safe enclosure, as it will attract rodents, snakes, crows, as well as larger predators.

During the day, turn birds loose only in an area with a predator-proof perimeter fence at least 5 feet high, preferably topped by electric wire. An electric scare wire outside the fence is also helpful to discourage dogs, coyotes, or foxes. Do not allow branches to overhang the fencing, providing climbing opportunities to small predators. Trees and fence posts can both allow raptors to perch.

Five-foot-tall fencing will keep most chickens in, although some can fly 6 to 7 feet. Clip flight feathers on their wings to prevent this.

Electrified netting can enclose chickens and exclude small predators. Grid systems of wire, monofilament, or Kevlar cord, plus netting, will exclude birds of prey from larger areas.

Pastured Poultry

Free-range poultry are more vulnerable to predators than are securely cooped birds, especially if they have to find their own shelter from predator attacks. If you are committed to pasture-raised poultry, several measures can reduce predation while balancing the cost of predator proofing:

- Use the best perimeter fencing you can or reinforce the existing fence.

- Keep the area mowed and cleared of brush to reduce hiding places for predators. Graze or keep other stock in a fenced free-range area — roosters, geese, sheep, goats, horses, or cattle will all provide some protection.

- Use livestock guardian animals (see page 257).

- Build small shelters for shade and places to hide from raptors.

- Use lengths of tightly strung, monofilament line and Mylar tape across open areas to reduce raptor predation.

- Music, antipredatory light systems, scare devices, security cameras, and baby monitors can help when predation threats continue at night.

Movable coops or chicken tractors, with or without protected forage pens, provide more protection and reduce predator familiarity with location. It can be harder to predator-proof chicken tractors on uneven ground. Fixed coops in the field can usually be made more secure at night either by closing the birds inside or by using electric mesh to surround the area. Netting to prevent baby birds from escaping or becoming trapped in the electric mesh is available.

LGDs and Poultry

Training LGDs to protect poultry safely presents a challenge because poultry are not their traditional animal to guard. They do not tend to bond to them as they do to stock. Supervision and time are required. An alternative is to allow LGDs to patrol around poultry coops, pens, or free-range areas. Some large producers have created alleyways around the entire free-range area for the dogs to patrol.

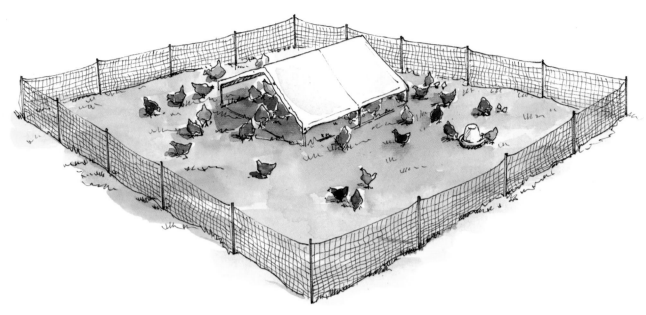

▲ Shelters offer pastured poultry protection from raptors.

▲ Portable electric mesh fencing can be placed around movable chicken tractors.
For added safety, secure the poultry inside the tractor at night.

Waterfowl

Ducks and geese can be kept in traditional coops, or adaptations can be made for providing water access for the birds by completely enclosing small pools or ponds, including a roof. Ducks and geese can also be fenced, although flying breeds may need their wings pinioned or flight feathers clipped. Electric wire on top and scare wire or buried mesh will keep out many predators. Adult weeder geese are often confined in 20- to 30-inch poultry netting. If waterfowl are kept free-range with access to ponds or waterways, the predator danger increases.

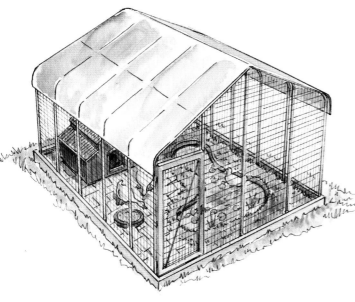

▲ Secure housing for waterfowl includes a predator-proof foundation, fencing, and a roof.

Rabbits

Outdoor hutches are healthier for rabbits or other small animals as long as they provide weather and predator protection. Hutches should be raised on legs or hung above the ground. Some rabbit raisers use LGDs to patrol the hutch area. Ensure that doors and latches are raccoon proof. Don't allow vegetation to grow under or next to the hutch. Placing the hutches inside a predator-proof fenced area provides increased protection.

Premade cages or hutches can be reinforced for predator protection. Although the holes in the bottom and sides of the hutch need to be large enough for feces to fall through, they often allow access for rats, weasels, opossums, raccoons, and other predators. Hardware cloth with ½-inch square holes can be stapled over these areas, but it may necessitate cleaning out the hutch more often.

Combining several hutches or building larger pens can house rabbit colonies or groups of rabbits living together. Concrete is the most secure flooring; however, many owners prefer dirt. To protect dirt colonies, perimeter fencing must run 3 feet underground to accommodate dens and tunnels and to prevent the rabbits from digging out or predators from digging in. The sides and top of the colony need to be predator-proofed as well. Larger colonies kept on pasture are harder to protect from aerial predators, although they can be fenced to prevent other predators. Truly free-roaming rabbits are at the greatest danger.

Rabbit tractors generally have floors constructed of wood or aluminum slats 1½ to 2 inches apart or wire mesh to prevent rabbits from digging out and yet allow for grazing. The sides need to be covered with hardware cloth. Rabbit tractors need to be heavy enough that predators can't tip them over. Some owners use LGDs in the area of rabbit tractors, or they bring the rabbits into a building or more secure pen at night.

Protecting Livestock and Poultry from Mountain Lions and Bears

To provide protection from these predators, animals need to be in fully enclosed housing. Free construction plans are available at mountainlion.org.

- Locate the housing away from trees, brush, or other cover.

- Place scare devices inside the area with predator-proof fencing.

- Construct walls with wood.

- Fasten the door with a sturdy bear-proof latch.

- Build the roof so it will support the weight of a mountain lion and a snow load.

- Cover all doors and windows with hardware cloth or chain link.

- Bury walls and pen fencing 1 foot into the ground.

- Construct fences with a substantial apron of fencing material, possibly several feet. Stake down and bury in 6 to 8 inches of dirt.

Upgrading an Existing Structure

- Check walls for soundness, reinforce and repair, cover all large gaps or openings. The entire surface can be covered with chain link if necessary. Animals need ventilation, but these areas should be covered with firmly attached chain link or hardware cloth.

- Extend the building and pen walls to the ground.

- Reinforce the door and roof.

- Build an apron around the outside of the building and pen.

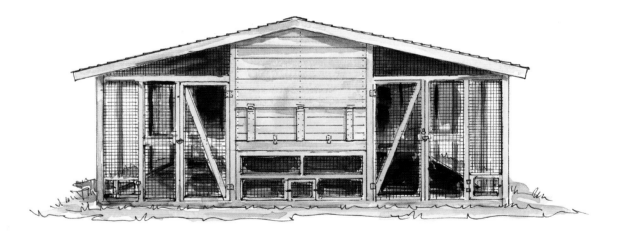

▲ Permanent poultry housing can be designed for both function and protection from even the largest predators, providing housing for separate groups of birds, easy collection of eggs, and daytime free-ranging.

Husbandry and Management of Grazing Animals

We know several key points about predator attacks on grazing animals:

- More predator attacks occur near water, in larger pastures, or in hilly or rough, brushy pastures.

- If we allow calving or lambing to occur in large pastures, leaving afterbirths and carcasses, it encourages more attacks.

- Predators are more likely to attack when natural prey is in short supply and when they have young to feed.

- Most predator losses are at night. Animals brought into night pens and confined during winter experience fewer losses.

- Predator-proof fencing for small farms is very effective, along with the use of night pens.

Thorough recordkeeping will give you insights into time and location of predator attacks on your property. Paddocks and areas closer to the farm center can be made more secure for birthing, nighttime, and winter confinement. Timed or synchronized breeding schedules can make this more manageable. In some cases, stock needs to be securely sheltered every night in predator-proof housing. Some producers move to indoor lambing sheds.

Rotate your pastures to accommodate predator pressure, and choose safer or sheltered areas for vulnerable times such as birthing. Move grazing away from wolf-occupied areas during calving or lambing, if possible. On very large pastures or on open range, create temporary and protected small areas for calving and lambing, especially at night. Remove dead, sick, or injured animals. Move weak lambs and calves and abandoned newborns to jugs or individual pens with their mothers or to more protected areas. Bury and compost afterbirth and stillborn animals, to keep domestic animals from developing a taste for them and to keep scavengers away. Groups of producers can work cooperatively to keep range areas free of carrion.

▲ Plastic protective collars may prevent a predator from biting or grasping the jaw area and under the neck.

▲ Indoor lambing sheds protect animals at their most vulnerable time.

Active Herding or Range Riding

The ancient technique of human interaction and supervision of grazing stock is still very valuable and has been proven effective against predators. Wolves and coyotes tend to stay away from areas with a human presence, even if it's only part-time presence. Either full-time shepherds working with their herding dogs and LGDs or range riders using horses or ATVs can deter predators and allow for the practice of good husbandry.

Humans can patrol at dawn and dusk when predators are active; check for predator signs such as tracks, scat, and upset animals; and even track wolves with radio collars to determine locations. People can support LGDs working with the stock by keeping the animals closer together for grazing rather than widely dispersed. Herders can move stock away from threatened areas as well as monitor range conditions to maintain the health of grazing land. On open range or large pastures, stock can be confined into more protected temporary pens of electric fencing or paddocks at night or at times of great vulnerability such as birthing.

Disruptive methods include visual, auditory, and other repellant techniques. Aversive methods work to condition a predator against engaging in a specific behavior. Both methods are more useful and successful when combined.

Disruptive Methods

Frightening and scare devices are old practices that can be adapted or improved today. A great value of these devices is that they can be used immediately after an attack or predator sighting or during more vulnerable times such as calving or lambing. The greatest disadvantage is **habituation**, which means the predator becomes used to the stimuli and begins to ignore them. Changing and moving the stimuli increase their usefulness.

▲ Range riders not only deter predators but also gain valuable information on their stock and provide a rapid response to new situations.

NIGHT LIGHTING

Night lighting is useful over corrals or pens, poultry housing, and around buildings. Light is effective in discouraging coyotes and other predators but not as effective with roaming dogs. Night lighting also allows an owner to check on stock with more ease. No proof supports the idea that strobe lights work better than regular lights; however, motion-activated lights are more effective than regular lights, as are novel or changeable patterns. Lighting can include flashing, rotating, regular, or search lights. Lights can be placed on timers for ease of use; that is, they will come on automatically even when you're not there. Lasers are generally used to deter pest and flocking birds rather than predators.

Pro: Inexpensive, provides night protection, and several can be placed in an area about 100 yards apart.

Con: Habituation, some raptors learn to avoid bright lights, can attract other raptors.

VISUAL DETERRENTS

Visual deterrents include Mylar tape and balloons, aluminum pans, and similar materials. Mylar tape moves, flashes, reflects sunlight, and can produce a clicking or humming noise. Mylar tape is available in different widths and can be hung as streamers or stretched between posts. Mylar or plastic balloons can be filled with helium so they float or filled with air and suspended. Some balloons are printed with large eyespots.

Pros: Inexpensive, widely available, movable, little or no noise, effective with birds of prey.

Cons: Predators can habituate within 3 to 5 days, not effective at night, less effective without wind; can be damaged by high wind.

◄ Nite-Guard lights use flashing red lights that mimic the eye shine of another predator and remain effective in fog and similar conditions. Fox Lights simulate the random pattern of flashlights. Lights like these are useful against nocturnal predators, including owls.

Fladry. A rope line of flags, pendants, or streamers, fladry is used to discourage birds of prey and wolves in particular. Flags moving in the wind add stimuli to an existing fence or function as a temporary fence or barrier against predators. They are very useful around calving or lambing grounds on open range, overnight pens on larger grazing lands, or grazing areas during periods of higher predator pressure.

The recommended installation is outside of the existing fence line on a separate rope line and posts. Since wolves are more likely to go under fladry than to jump it, the bottom of the flags should be 6 to 8 inches off the ground. The rope line itself is generally 15 to 24 inches high. The line needs to be taut, a consistent height, and avoid brush. At times, existing fence, trees, or natural objects can be incorporated into a fladry line. Larger flags or fabric panels may be more effective. Wolf vision is very motion sensitive, but wolves do not see the same range of colors as humans see. Red-, orange-, and gray-colored flags have been tested, but color may not be important. Plastic caution tape or nylon pieces and mason line can be used to create a homemade fladry. Nylon is more durable, and stock may chew the plastic.

Fladry is still under development, and improvements should continue.

Turbo or electrified fladry is 2 to 10 times more effective than regular fladry, with a longer period before habituation. It also provides protection at night. The cost is approximately twice that of regular fladry. Installation requires greater care to avoid touching ground, brush, or objects that will ground out the fence. Turbo fladry is especially suited to smaller pastures, overnight penning, and calving or lambing pastures.

Pro: Inexpensive, effective, movable, no noise. Fladry works well against wolves for up to 2 months if installed correctly.

Con: Not as effective at night unless electrified, habituation for birds after 3 to 5 days, setup time, periodic maintenance when installed for longer periods. High winds will tangle fladry or wrap the flags around the rope or cable. Not proven effective against coyotes, bears, mountain lions, or wolves in some situations.

▲ Commercial fladry is not yet widely available; fortunately, homemade fladry is also very effective. The fladry rope should be strung as tightly as possible.

Scarecrows and effigies of owls or vultures.

Scarecrows can be static, mechanical (they pop up and move), or inflatable with noise and light cycles. Scarecrows that twist in the wind, move their arms, and wear human clothing can be effective for up to 3 weeks, and then you'll need to move them elsewhere.

Plastic effigies of owls can be static, or the wind can turn the head. Research has shown that effigies of dead vultures hung where damage is occurring will repel black and turkey vultures.

Pro: Inexpensive, can be movable, can be noise-free.

Con: Static items must be moved every 2 to 3 days to prevent habituation, not as effective at night, only protects small area.

AUDITORY

Music, especially with bass or talk radio, has unpredictable rhythms and sounds like human voices. Recorded sounds include alarm or distress calls, dogs barking, gunfire, human screams, music, predator calls, or synthesized sounds. Recorded sounds can be species-specific. Some commercial sonic units are programmable and include lights, and additional speakers can extend coverage.

Horns, sirens, and propane cannons should be set to go off every 8 to 10 minutes or at random intervals, and they should be moved every 3 to 4 days. Gunfire and fireworks are labor-intensive, temporary, and best used only in response to the immediate presence of a predator.

Using bells on livestock is a traditional method of noise deterrence. One study asserts that coyotes and bears do not attack the belled animals in a flock or herd. Some bear specialists assert that bells now attract bears, and they discourage hikers from wearing bear bells. Predators are likely to habituate to belled stock but if combined with other aversive methods, such as LGDs, bells may condition predators to avoid a particular group.

Pro: Useful day or night, delayed habituation, some methods inexpensive.

Con: Noise may disturb neighbors; some methods expensive. Prohibited in some cases.

◄ The traditional practice of belling some members of a flock or herd still has merit, especially if the predator also encounters other disruptive or aversive responses to attacking belled stock. The loud noise made by a belled animal in flight may also discourage an attack.

REPELLENTS

No proven carnivore repellents are available, although several compounds are being tested. The following are some traditional methods:

- Both coyotes and dogs may be repelled by the chemical **pulegone**, found in mint.

- Ammonia can repel skunks and other small predators.

- Predator urine may discourage some animals. Use caution when handling any animal urine, and do not use it around garden plants.

- Female raccoons can be repelled by biological products based on the male raccoon's gland secretions, which she interprets as a threat to her offspring.

- Bears have been discouraged from beehives with **thiabendazole** (TBZ).

- Many LGD owners believe urine marking by their dogs helps to discourage predators.

Large Predator Warning System

Various wildlife agencies are testing combination systems in specific situations.

RAG (radio-activated guard)

This system uses lights and sound activated by radio-collared wolves, bears, or mountain lions. Tracking these collared animals can also alert owners to the nearby presence of a major predator so that extra precautions can be taken.

MAG (movement-activated guard)

This system uses an infrared detector with light and sound response.

Aversive Methods

Aversive methods condition a specific predator not to perform a behavior, such as stalking or attacking poultry or livestock. Aversive techniques include the use of hazing, multispecies grazing, livestock guardian animals, hard release of problem animals, and bear shepherding with Karelian bear dogs (page 259).

HAZING

The act of disturbing an animal until it leaves, hazing employs multisensory techniques — visual and auditory — as well as threatening or aversive behaviors. While the public is encouraged to haze coyotes, the hazing of bears, mountain lions, or wolves can be more dangerous and is often left to experts, although individuals should employ these techniques if confronted by an animal. See Coyote Hazing Tips (page 65).

MULTISPECIES GRAZING

The use of guardian llamas and donkeys is basically a form of multispecies grazing. Other animals such as cattle or horses can also offer protection to smaller stock, although often the different species are not inclined to graze together.

The term **flerd** describes a multispecies group that has bonded together through penning for 30 to 60 days. Horned cattle can be especially effective in a flerd. Consider identifying and incorporating a few of the more aggressive and protective ewes or cows, even if they are more difficult to handle or less productive. Rams, bucks, and bulls are naturally more protective and aggressive. Even one or two larger animals will help protect smaller animals, such as sheep or goats.

Livestock Guardian Animals

Livestock guardian animals offer many advantages. They are very alert to potential threats. Predators do not become habituated to their responsive attacks, which are both disruptive and aversive. The use of livestock guardians is also considered a predatory-friendly marketing approach.

In sheep operations, the use of livestock guardian animals is the second most common non-lethal method of predator control, after fencing. Approximately 37 percent of sheep raisers use guardian animals, with LGDs the most popular choice (23.5 percent), followed by donkeys (8.2 percent), and llamas (5.4 percent). The use of both LGDs and donkeys has more than doubled in the last decade, as their value has been proven.

LIVESTOCK GUARDIAN DOGS

LGDs are a very old partnership dedicated to protecting sheep, goats, or other stock. Selected over hundreds of years, these breeds have a low or non-existent prey drive, a longer period of social bonding to stock, and a physical appearance that suggests friend not wolf to animals. LGDs possess instinctive responses and behaviors to threats. The specific breeds of LGDs were developed over an area from southern Europe through Central Asia — specially to protect sheep and goats from large predators. The use of LGDs has been shown to reduce predation by 60 to 70 percent or more.

Shepherds lived and worked together with their LGDs in their homelands. All LGDs are raised with human or older dog supervision. It is a myth that livestock guardian animals should be raised with minimal human contact. While they need extensive socialization to stock, all livestock guardians need to be safe to handle and care for. Without regular handling, guardians will become uncontrollable and nearly feral.

Research has shown that LGDs are more effective when farmers or ranchers work directly with the dogs and do not leave them alone to guard stock. Dogs are the least successful when asked to guard stock dispersed over a large area, in very rough or wooded areas, and where owners have spent only minimal time with the dogs. Ideally, LGDs should be established with their stock before peak predator times and before the predators establish a territory. Wolf attacks are more successful when dogs are outnumbered, when the dogs are placed in an already established wolf territory, or when people are not present to provide support during an attack. A single dog is more likely to accept or tolerate a wolf than packs of dogs will be. Most important, livestock or poultry owners need to be proactive and not wait until after an attack to begin looking for a fully trained adult dog. Using LGDs successfully requires the commitment to obtain, raise, and train good dogs as insurance against future needs.

LGDs patrol and mark their territory. They are active at night during the time of greatest predation and will bark to warn off potential threats and to alert their owner. The dogs are self-thinkers and will work cooperatively in pairs or packs to both defend the stock and charge the predator. LGDs employ a graduated response to predators — barking, posturing, and charging usually drives the threat away — and they attack only when the predator does not leave. LGDs can be used in situations with the most dangerous predators. Both females and males are equally effective.

The most important factors in achieving an effective livestock guardian include choosing a dog from working parents who are both recognized LGD breeds, providing good socialization to stock, and giving positive training and supervision until the dogs are reliable. Guarding poultry is the most difficult task for LGDs and not a traditional one in their homelands. Guarding equines can also be problematic if the horses or donkeys have a strong antipathy to dogs.

▲ LGDs have been protecting livestock for more than 2,000 years. In their homelands, LGD pups are never left alone with stock, but are carefully supervised and trained by shepherds and older, experienced dogs.

Advantages

- Effective against large and small predators, as well as raptors, small reptiles, and feral animals such as dogs, cats, and hogs.

- Works well and effectively in pairs providing more protection. Can work in LGD packs, with diversified roles, to combat the largest predators or packs.

- Facilitates use of pastures with heavy predator pressure and night grazing.

- Reduces need for human labor for night penning, patrolling, and so forth.

- Can be used, especially in pairs or groups, with very large flocks on large or rough pastures or open range. Dogs are more effective than other livestock guardians on open range.

- Guards sheep, goats, cattle with calves, llamas, alpacas, miniature horses or donkeys, and poultry.

- Often nurturing to young stock.

- Predators do not become accustomed or habituated to LGDs, enabling long-term use. LGDs actively patrol and mark pastures. LGDs are most active at night during time of greatest predator activity. LGDs also provide hazing of predators.

- Alerts owners to disturbances; provides protection for family and farm from predators.

- Provides a graduated response to threats, from barking to charging or bluffing. Will physically attack if predator not discouraged.

- Meets Predator Friendly® guidelines (see page 270).

Karelian Bear Dogs

Karelian bear dogs (KBDs) are hunting and watch-dogs originating in a large area of the northern latitudes from Scandinavia into Siberia. They have demonstrated significant value in deterring grizzly bears and other large predators such as mountain lions. Since the 1990s, the nonprofit Wind River Bear Institute, founded by biologist Carrie Hunt, began training handlers and KBDs for both black and grizzly bear conservation and other nonlethal predator control efforts. Karelians are trained for use with bear management specialists, trained volunteers, and people in bear country. KBDs have three essential roles: *bear conflict dogs* for management specialists, rangers, or wardens; *bear protection dogs* for biologists, ranchers, or outfitters; and *companion dogs* for hikers and others.

Trained KBDs and handlers are used in some US and Canadian national parks and in several western states or provinces. KBDs are used to push or shepherd bears out of human conflict areas, both urban and recreational, including camping areas and residences. The dogs are used in conjunction with aversive conditioning techniques such as bells, firecrackers, and pepper spray in structured experiences to teach bears to recognize, fear, and avoid human areas. This conditioning is often used in a "hard release" of a problem animal. KBDs are also used to locate, track, and assist with removal of problem or injured animals.

KBDs bond closely to their owners yet can be quite independent in their decision-making. Karelians have a high prey drive and a strong desire to chase. Owners report that they easily jump 6-foot fences and will do so in order to hunt. Karelians are strongly territorial, protective, and dog aggressive. When working they are courageous, intense, and focused. They are not generally well suited as a pet, especially in urban environments.

▲ Not LGDs, Karelian bear dogs are used to haze bears away from human conflict areas.

Considerations

- Requires excellent fencing unless on open range, otherwise they will roam, increase their patrol area, or chase predators over a very large area.

- Requires socialization, training, and supervision until approximately age 2; otherwise may harass or injure stock.

- LGDs are independent self-thinkers and may not obey directions.

- Dogs that are not handled regularly can become nearly feral.

- Can be highly protective of stock and defensively aggressive toward strangers.

- May not tolerate other dogs, although many become accustomed to familiar herding dogs.

- Cost of dog food, medications, veterinary care, and initial purchase price. Expected working life 12 to 14 years.

- Will bark, especially at night, and may dig dens.

GUARDIAN DONKEYS

Donkeys are used less frequently than dogs as guardians. Donkeys are very territorial and strongly aggressive to canines, lending their protection to the animals they live with. They are very alert to potential threats. Donkeys may bray loudly at a disturbance, followed by chasing a canine. If they engage in an attack they will kick, bite, and slash at the intruder. Stock may also see the donkey as a protector and gather near it.

Donkeys are less social than llamas and require a longer period of social bonding with potential charges. A foal raised with sheep or goats is more likely to live well with stock. A jenny with foal is the most protective. Either mature jennies or neutered males can be good guardians, while intact males can be dangerous to stock and humans. Guardian donkeys should be standard sized or larger, since miniature donkeys are too small to deal with many North American predators.

Livestock Guardian Dog Breeds

LGDs comprise a specific set of breeds, not a job. Only livestock guardian breeds and crosses between these breeds possess the proper instinctual behaviors, protectiveness, nurturing nature, size, and coat to work as full-time livestock guardians. You'll notice some size, coat, and behavioral differences among breeds, and some breeds are more reactive or defensively aggressive while others are more tolerant of strangers. Crossbred LGDs are also effective, but only if both parents are from recognized LGD breeds. The following breeds are available in North America.

- Akbash Dog
- Anatolian Shepherd Dog
- Armenian Gampr
- Cao de Gado Transmontano
- Caucasian Ovcharka
- Central Asian Shepherd
- Estrela Mountain Dog

- Great Pyrenees
- Kangal Dog
- Karakachan Dog
- Komondor
- Kuvasz
- Maremma Sheepdog
- Polish Tatra Sheepdog

- Pyrenean Mastiff
- Sarplaniac
- Slovensky Cuvac
- Spanish Mastiff
- Tibetan Mastiff
- Tornjak

▲ Guardian donkeys should be standard sized or larger. Miniature donkeys are inappropriate to use as a guardian against predators.

Advantages
- Generally effective against small canines and perhaps bobcats.

- Best suited to flocks of fewer than 100 animals in fenced pastures without rough terrain or heavy brush or trees.

- Able to guard sheep, goats, calves, deer, or poultry.

- Easy to feed and fence.

- Long working life.

- Bonds to stock without extensive socialization or training.

- Less disturbance to neighbors unless individual donkey brays frequently.

Considerations
- Cannot protect against more than one roaming dog or coyote. Vulnerable to multiple dogs or coyotes, bears, mountain lions, and wolves.

- Does not provide protection against most small predators, feral animals, or large birds.

- May injure stock.

- Can be dangerous to humans and difficult to handle if not socialized and trained.

- May not tolerate herding or farm dogs.

- As a solitary animal, may not bond to stock.

- May not exhibit guardian behavior against canines.

- May ignore threats to stock or injure animals, especially during breeding and birthing periods.

- Often do not work well in pairs, since a single donkey will bond better to stock.

- Do not have the protective undercoat of horses and will need shelter from rain or snow.

- Ruminant feeds containing *coccidiostats* are poisonous to donkeys.

GUARDIAN LLAMAS

Llamas are very social. A single llama will tend to associate with other animals and be protective of them. Dominant males naturally guarded groups of females with young. Some llamas will position themselves between the threat and their flock or will attempt to herd them away. They often place themselves on higher ground to scan for predators and are reliably aggressive to canines. Llamas usually alert to a threat by a high-pitched call, followed by posturing or spitting, followed by kicking or pawing at the attacker.

Llamas may be a good choice for some situations, but they are unable to confront more serious predators. Use only mature adult females or neutered adult males. In North America, alpacas are not appropriate guardians because of their diminutive size against the size of our predators.

As social herd animals, guardian llamas will act protectively in defense of their flock mates.

Advantages

- Generally aggressive and effective against small canines, such as foxes, coyotes, or dogs.

- Able to guard sheep, goats, cattle with calves, deer, alpacas, or poultry.

- Easy to feed and fence with stock.

- Best suited to flocks of less than 100 animals in fenced pastures without dense vegetation and close to the farmstead.

- Long working life.

- Naturally social to pasture mates and bond readily to stock.

- Creates little disturbance or threat to neighbors.

Considerations

- Cannot protect against more than one roaming dog or coyote.

- Vulnerable to multiple dogs or coyotes and bears, mountain lions, or wolves.

- Does not provide protection against most small predators, feral animals, or raptors.

- May injure or attempt to breed stock, especially intact males.

- May injure working farm dogs.

- If not socialized and trained, can be dangerous to humans and difficult to handle.

- Often do not work well in pairs, since a single llama will bond better to stock.

- Not suited to hot, humid weather.

Predator Protection for Home and Yard

The following precautions are useful in all situations, although householders in large-predator habitat need to be extra proactive and careful. Be aware of habituated coyotes, bears, mountain lions, and wolves in your area. Practice recommended hazing techniques around your home and neighborhood. Eliminate shelter, water, and food for both predators and prey. Remember that your yard is a habitat.

Fencing

- Where predators are an issue, fence securely and/or improve existing fencing. The presence of large predators means you need serious exclusion fencing (see pages 241–245).

Landscaping

- Clear brush to improve visibility and sight lines and eliminate potential cover.

- Trim tree branches that could allow a predator to climb or drop into your yard.

- Remove rubbish, lumber, firewood, and rock piles from areas next to fences and buildings.

- Store wood and building materials 18 inches off the ground and 1 foot from the building walls.

- Store firewood in covered and latched bins.

- Trim vegetation near buildings and foundations to eliminate cover for rodents and small predators.

- Reduce open water sources in arid areas.

- Be especially careful of children's play areas and walkways used at night.

Plants That Attract Trouble

Landscape plants that attract deer and/or coyotes vary by area, but some common culprits include:

Deer: snow, service, and elderberry shrubs; cherries, plums, persimmons, strawberries; arborvitae, European mountain ash, rhododendrons, azalea, sea holly, yews; tulips, hardy geranium, candy lily, winter creeper, clovers, and many others

Coyotes: strawberries and other berries, watermelons, sarsaparilla, balsam fir, white cedar

Feeding Animals

- Feed pets indoors, or do not leave food out overnight.

- Equip bird feeders with baffles or guards and spill pans. Clean up the ground underneath.

- Hang suet feeders 10 feet high in mountain lion or bear country.

- Do not feed feral cats, squirrels, raccoons, or any other wild animal.

- Do not use landscape plants that deer or coyotes feed on (see box above).

- Securely house backyard poultry, waterfowl, pet birds, and rabbits to discourage predators, especially at night.

Food Waste

- Dispose of all garbage and pet food in securely covered containers.

- Clean outdoor kitchens, grills, and fish-cleaning stations.

- Compost properly by covering all food scraps and avoiding meat scraps. Use commercial composters, mesh-covered frames, or 55-gallon drums for the best protection.

- Remove ripe fruit and vegetable crops.

- Clean up rotting fruit, nuts, and seeds on ground.

Access

- Secure pet doors at night or use electronic openers.

- Shut garage doors and other doors at night.

- Repair or cover all openings under decks, porches, window wells, basements, foundations, and outbuildings.

- Cover small openings, overhangs, and vents with ½-inch hardware cloth. Very small gaps can be filled with pliable wire mesh.

- Cover chimneys with safe, commercially designed caps. Metal or plastic flashing and plastic or metal spikes can protect animals from climbing trees, poles, and so forth.

Lighting

- Parking and walkways should be brightly lit when in use.

- Motion-sensitive lighting may be useful although animals may habituate to it.

Prevention

- Be alert to the presence of predators — animal tracks, burrows, dens, tunnels, scat, and other signs. Fill old burrows or tunnels with gravel or cover with mesh.

- Eliminate rodent infestations.

- Wires, Mylar tape, balloons, flags, reflective objects, and other scare devices can discourage birds of prey, ravens, crows, and magpies.

- Variable light and sound guards and motion-sensitive sprinklers may be useful to discourage some predators and birds.

Tactics That Don't Work Well against Predators

- If you fill a hole, the animal is likely to reopen it.

- Flooding dens or tunnels with water is temporary but may drive an animal out if it doesn't drown.

- Live traps require the animal to be relocated, but to where? Don't take a trapped animal down the road to bother another farmer.

- Mothballs work in small areas but are highly flammable and possibly carcinogenic.

- Sticky substances can catch birds or other small animals, but then the animal must be killed humanely or it expires slowly.

- Ultrasound devices generally don't work because most mammals can't hear them, they rapidly lose intensity in a few feet, and they don't penetrate walls or other insulated materials, including paper or cardboard. They are often illegal to use in bat areas.

- Electromagnetic devices have been proven completely ineffective.

- Diversionary feeding is sometimes attempted to draw predators away from stock; however, it may condition predators to expect handouts and may ultimately increase populations.

Safety

- Do not confront or corner a predator, regardless of size.

- Take children and pets indoors in the presence of a predator. Most predators will leave on their own.

- Handle scat and dead animals properly to prevent exposure to disease or parasites. Clean and disinfect areas that have been contaminated.

Predator Protection for Pets

Cats and small- to medium-sized dogs are the most vulnerable, as are other backyard pets or poultry.

- Feed animals indoors if possible. If animals need to be fed outdoors, do not leave food out at night, clean up spilled food, and secure stored feed inside buildings. Pet and wild bird food will attract the prey of both small and large predators, which in turn attract the predator.

- Do not leave open water outside at night in arid climates, since it will attract both prey and predators.

- Keep pets indoors from dusk to dawn. Most predators are not active during daylight hours.

- Keep pets in fenced areas. Roaming pets are at the greatest risk of an attack, especially at night.

- If walking a dog after dark, keep it on a leash, stay in a brightly lit area that is free of potential cover for a predator, and make noise.

- Situate animal housing away from woods or heavy brush areas.

Dogs

Never chain or tether a dog, leaving it vulnerable to predators. A doghouse alone is not a safe refuge from a predator. If a dog is left alone outside, it needs to be enclosed in a well-fenced yard or in a completely enclosed kennel or run.

▲ Where predation on small- to medium-sized pets is a threat, covered runs or pens and a safe retreat indoors are the safest option.

Build the kennel of well-attached chain link, heavy wire mesh, or livestock panels. Lay a livestock panel or wire mesh under dirt or gravel areas to prevent your dog from digging out and a predator from digging in. Cover the entire top of the kennel with chain link or panels, not a tarp, if there is any danger of predators climbing up and into the kennel area.

A doghouse inside the kennel can provide a place to hide, but the best solution would be an access door into a building or the house. Electronic or magnetic doors are available that only open for a small device on a pet's collar. Lock this door at other times.

Cats

Pet cats can be allowed safely outdoors in securely enclosed porches, pens, and cat runs. Some owners construct a safety cat post made of climbable material (such as sisal rope), 7 feet high with a flat platform on the top.

To prevent your cat from climbing out over the fence, install a U-shaped fence topper or overhang out of small-gauge mesh, at least 2.5 inches square. Also use the mesh to cover any openings or as an apron to prevent digging. Tree guards or baffles are also useful to prevent climbing. This fence overhang may keep your cat in, but it won't prevent other cats or other predators from getting into your yard.

Beehives

In addition to the threats of disease, mites, and other pests, beehives are highly attractive to some animal predators. Bears are the major threat, capable of causing serious destruction and significant economic loss. A clean, well-tended bee yard will discourage predators and reduce temptations; however, good fencing can exclude raccoons, skunks, and bears. Strapping the hive together and weighing down the top will also make it harder for a predator to take it apart or gain access.

Mice. Install mouse guards to prevent mice from entering the hive, especially before winter. Wooden reducers can be enlarged by chewing mice.

Raccoons. Place a heavy rock, brick, or cement block on top of the hive to prevent raccoons from removing top boards.

Skunks. Raise hives higher than a skunk can reach, up to 3 feet. Place plywood with a nail "pincushion" in front of the hives. Predator-proof fencing, chicken wire, and netting will keep skunks away from hives. Bury wire netting, mesh, or hardware cloth 6 inches down and extending out 6 inches.

Bears. Place electric netting or fencing around the hives, and use sturdy livestock panels for the best combination of deterrents.

▲ Backyard beekeepers can prevent predation by grouping their hives together, surrounding them with closely spaced electric fencing, and weighing down the hive top.

Moving Forward Together

Coexistence means to live together, or close to one another, in relative peace, or at least in respectful tolerance, despite differences and conflicts. Not only do we need to learn how to coexist with predators but the various stakeholders and communities that are dealing with predators also need to work together constructively rather than occupy entrenched positions.

Predator control, especially large predator control, inspires strong and impassioned responses on different sides of the issue. We need to acknowledge these different opinions, beliefs, and points of view: preservation or conservation groups and producers; wildlife researchers and wildlife managers; state and federal agencies; eastern and western regions; urban, suburban, exurban, and rural residents — all of whom are now finding predators, large or small, in their backyards.

The concept of coexistence implies respecting differences and resolving conflicts, and this means compromise and cooperation. People can come together by listening openly to one another and then working on collaborative efforts. Groups that include particular interests can work together to agree on a mission and to develop specific workable projects.

Reducing Conflict over Predators

Since the return of the large predators, and the expansion of smaller predators deep into urban areas, the learning curve on how to deal with them has been rapid and sharp. Farmers and ranchers must adapt to the presence of coyotes, bears, wolves, and mountain lions in areas where they haven't existed for decades or where they are now protected.

Livestock raisers have been hit hard, but an increasingly hot spot has also occurred at the intersections of rural and suburban living where coyotes and mountain lions now live among homes and roam recreation land. Joggers and dog walkers now must know how to react in the presence of a predator, a skill once needed primarily in the backcountry. Even backyard poultry raisers with lovingly constructed coops have been forced to learn about raccoon, weasel, and owl predation.

Lethal responses to predation, while sometimes necessary, are generally not effective unless they are highly selective and targeted. Research has shown that when adult breeding wolves are killed, the pack and its territory are destabilized, which can lead to increased hunting of livestock for food and increased reproduction. Research also shows that you can't kill enough coyotes each year to make a difference in their population — and that the new coyotes who move into the territory may be worse. In addition, many people kill any snake, small predator, or perceived predator in their yard, which is an unnecessary overreaction.

Coexistence is possible when science, research, and practical experience help us minimize conflicts. Successful use of nonlethal methods requires a more complete understanding of predator behaviors, a shedding of preconceptions and myths, and a flexibility to react to changing needs or situations. Ongoing research is needed on livestock predation patterns and their causes, effective husbandry techniques and deterrents, and their uses in combinations and at different times or situations.

We also need answers to other questions such as these.

- Can we accommodate large-predator movement through residential and other land developments by creating **carnivore ways** or safe corridors to mitigate fragmentation of habitats?

- How do we best provide for public safety while supporting the presence of predators?

- Are there additional practices that we can adopt to avoid or minimize predator conflicts in urban and suburban areas?

- What are the economic impacts of nonlethal methods of poultry and livestock protection?

- Are compensation programs the best way to address the inevitable predation that occurs?

As we learn more, we must continually reevaluate our approaches and tools.

Supporting Producers

Much of the public is far removed from farming or ranching life. The sheer violence of a predator attack on domestic animals can be deeply disturbing, in addition to the financial loss from predation. It is also very difficult to account for all predation deaths, leaving many producers to believe the public is underestimating their losses. Livestock producers correctly point out that the rest of the population depends on the food they provide. The public can support good husbandry methods by buying products from Wildlife Friendly® or Predator Friendly® certified producers (see page 270).

Livestock producers, both small and large, need to plan ahead for potential predation, researching and implementing appropriate practices in advance of an emergency situation. Organizations like Defenders of Wildlife can provide financial assistance for the adoption of best practices including: a human presence or range riders with animals; the use of livestock guardians; portable fencing or fladry protection for night penning of vulnerable animals; bear-resistant trash cans, the use of safe grazing areas during hazardous periods; or predator research. Organizations also work to sponsor community education and outreach.

Wildlife managers at both state and federal level need to involve all of the stakeholders in discussion and problem-solving issues, reevaluate their processes for decision-making, and demonstrate both compassion and flexibility in implementation of their programs and regulations. Collaborative management programs and efforts such as the Interagency Grizzly Bear Committee or the Mexican Wolf/Livestock Coexistence Council enable this kind of communication and cooperative work. Funding state wildlife work through hunting fees leaves those organizations dependent on protecting game animals and viewing sport hunting of predators as a primary management tool — other funding avenues could shift future discussions and planning.

If You Encounter a Working LGD

LGDs are one of the most effective means of non-lethal predator control. With their increasing use on private and public land, encounters between the dogs and visiting humans occur more often. LGDs at work take their job seriously. They generally bark and bluff to warn off threats, and that might include you. If they believe their charges are threatened, they will escalate. Their owners want you to be safe, but you need to behave appropriately around these serious working dogs. Knowing how to react is essential whether you are visiting a farm or hiking on public land.

Behind a fence

- Don't attempt to pet or feed the dogs.

- Don't throw things at the dogs or verbally harass them in a threatening manner.

- Don't open gates or enter the area without the owners.

Hiking or biking on open land

- Look for informative signs that LGDs are present.

- Keep pet dogs on a leash so they do not chase sheep. LGDs are more disturbed by hikers with dogs. It is not advisable to take a pet into an active grazing area.

- If you observe a flock, do not approach the flock or the LGDs.

- Stay at a distance and alter your path to move widely around the sheep and not through them.

- Dismount from your bike and walk it. Keep it between you and the animals.

- Keep calm and do not harass the dogs. Speak softly and quietly to the dogs if they come near. Do not wave a stick at the dogs or yell at them.

- Turn around and walk back the way you came if you are uncertain or anxious.

Education and Outreach

Educational efforts inform the public on local predation threats and explain how to use control methods and techniques successfully. Many of these efforts are conducted with sheep, goat, and cattle producer groups, but they are also essential for suburban or urban homeowners to learn the best methods for discouraging conflicts with wildlife, landscaping intelligently, and eliminating attractants. Communities can implement practices that discourage predator habituation, which often leads to their destruction. For success, these measures must be practiced across a community.

We must all learn how to interact safely with both small and large predators, not to fear them but to be prepared with proper responses and precautions as with any activity. Reckless or uninformed actions are linked to many predator attacks.

The public also needs to respect the nonlethal predator control efforts of farmers and ranchers, such as how to behave appropriately around livestock guardian dogs (LGDs) with stock. Complaints about barking or demands for their removal from public grazing land do not support nonlethal methods of predator control. While public lands and wildlife should be preserved for everyone, visitors, hikers, and cyclists need to pay attention to signs and warnings from authorities and not behave recklessly when they enter wildlife areas.

Predator Friendly® and Wildlife Friendly® Farming

Consumers are increasingly aware of positive environmental efforts and more humane alternatives for raising meat, eggs, and other products. Cage-free egg marketing is one example of successful added value for humanely raised products.

Originally called Predator Friendly, this marketing program is now part of the **Wildlife Friendly Enterprise Network,** www.wildlifefriendly.org, which certifies agricultural businesses that support the coexistence of people and nature and the conservation of biodiversity and ecosystems. Certification allows farmers, ranchers, herders, harvesters, artisans, and businesses to use the logo to market their products to individuals and commercial markets such as restaurants, grocery stores, and web-based retail sites.

Owners of Predator and Wildlife Friendly farms and ranches, or who use grazing leases on public or private land, can seek certification through meeting established standards. The standards help demonstrate that the farms provide and conserve habitat and corridors for native predators and prey, while implementing nonlethal management practices that protect poultry or livestock and eliminate attractants.

See page 240 for techniques that fit the Wildlife Friendly Network's guidelines.

Products That May Be Labeled Predator Friendly

- **Wool:** raw wool, felted wool, yarn or thread, woolen clothing, piece goods

- **Animal fiber, yarn, and goods:** Angora rabbits, Angora goats, alpaca, guanaco, llama, vicuna, camel, yak, bison, donkey, dog

- **Leather goods:** sheepskins, cowhide, emu, ostrich

- **Meats:** beef, bison, chicken, duck, emu, game birds, goose, lamb, ostrich, pork, turkey

- **Dairy products:** butter, cheese, milk, sour cream, yogurt

- **Other:** eggs, honey, soaps and lotions from animal by-products, breeding stock

Appendix

Predator Control Guides for Home, Recreation, Farm, and Ranch

Predator Control for Sustainable and Organic Livestock Production. ATTRA Publication #IP196, 2002.

Gese, Eric M. *Lines of Defense: Coping with Predators in the Rocky Mountain Region*. Utah State University Extension. www.aphis.usda.gov/wildlife_damage/protecting_livestock/downloads/predators_booklet7.pdf, 2005.

Felidae Conservation Fund. A Rancher's and Farmer's Guide to Keeping Livestock Safe from Mountain Lions. www.felidaefund.org, 2011.

Lance, Nathan, Steve Primm, and Kristine Inman. *Wolves on the Landscape: A Hands-on Resource Guide to Reduce Depredations*. www.fwp.mt.gov/fwpDoc.html?id=69188, nd.

Living in Bear Country: Guidelines for Protecting People, Property and Bears. Defenders of Wildlife. http://www.defenders.org/sites/default/files/publications/living-in-bear-country-guidelines -for-protecting-people-property-and-bears.pdf, 2008.

Sowka, Patricia A. *Living with Predators Resource Guide: Recreating in Bear, Wolf, and Mountain Lion Country*. Living with Wildlife Federation, 2013.

Stone, Suzanne Asha. *Livestock and Wolves: A Guide to Nonlethal Tools and Methods to Reduce Conflicts*. Defenders of Wildlife, 2008. http://www.defenders .org/sites/default/files/publications/livestock_and _wolves.pdf.

Texas Sheep and Goat Predator Management Board. *Predator Control as a Tool in Wildlife Management*, 2014.

Predator Management for Small and Backyard Poultry Flocks. 2014. articles.extension.org/pages/71204 /predator-management-for-small-and -backyard-poultry-flocks.

Poultry Predator Identification: A Guide to Tracks and Sign. www.ouroneacrefarm.com/poultry-predator -identification-a-guide-to-tracks-and-sign, 2015.

Mountain Lion Foundation Secure Livestock Enclosures in Mountain Lion Country. mountainlion.org /portalprotectsecureenclosures.asp, 2017.

Organizations Providing Additional Resources for Predator Control

The Conservation Fund
www.conservationfund.org

Coyote Lives in Maine
www.coyotelivesinmaine.com

Coyote Watch Canada
www.coyotewatchcanada.com

Defenders of Wildlife
www.defenders.org

Farming with Carnivores Network
www.farmingwithcarnivoresnetwork.com

Florida Panther Net
www.floridapanthernet.org

Get Bear Smart Society
www.bearsmart.com

Interagency Grizzly Bear Committee
www.igbconline.org

International Wolf Center
www.wolf.org

Living with Wildlife Foundation
www.lwwf.org

Mexican Wolf/Livestock Coexistence Council
www.coexistencecouncil.org

Mountain Lion Foundation
www.mountainlion.org

People and Carnivores
www.peopleandcarnivores.org

Project Coyote
www.projectcoyote.org

Urban Coyote Research
www.urbancoyoteresearch.com

US Fish and Wildlife Service
www.fws.gov

Western Wildlife Outreach
www.westernwildlife.org

Bibliography

Avery, M. L., and J. L. Cummings. *Livestock Depredations by Black Vultures and Golden Eagles*. USDA National Wildlife Research Center, 2004.

Baron, David. *The Beast in the Garden: A Modern Parable of Man and Nature*. New York: W. W. Norton, 2004.

Beeland, T. Delene. *The Secret World of Red Wolves: The Fight to Save North America's Other Wolf*. Chapel Hill: University of North Carolina Press, 2013.

Bruskotter, Jeremy T. *The Challenge of Conserving Carnivores in the American West*. Ohio State. Accessed at faculty.nelson.wisc.edu/treves/pubs/Bruskotter_Treves_Way-inpress.pdf.

Busch, Robert H. *The Wolf Almanac: A Celebration of Wolves and Their World*. Guilford, CT: Lyons Press, 2007.

Butz, Bob. *Beast of Never, Cat of God: The Search for the Eastern Puma*. Guilford, CT: Lyons Press, 2005.

Cartano, Carol. *Myths & Truths about Coyotes: What You Need to Know about America's Most Misunderstood Predator*. Birmingham, AL: Menasha Ridge Press, 2011.

Chadwick, Douglas H. *The Wolverine Way*. Ventura, CA: Patagonia, 2010.

Chadwick, Douglas. "Wolf Wars." *National Geographic*. March 2010.

Clark, Susan G., and Murray B. Rutherford, eds. *Large Carnivore Conservation: Integrating Science and Policy in the North American West*. Chicago: University of Chicago Press, 2014.

Dohner, Janet Vorwald. *Livestock Guardians: Using Dogs, Donkeys, and Llamas to Protect Your Herd*. North Adams, MA: Storey Publishing, 2007.

Dolbeer, Richard A. "Identification and Assessment of Wildlife Damage: An Overview." *Prevention and Control of Wildlife Damage*. Washington, DC: United States Department of Agriculture, 1994.

Elbroch, Mark. *Mammal Tracks & Sign*. Mechanicsburg, PA: Stackpole Books, 2003.

Eisenberg, Christina. *The Carnivore Way: Coexisting with and Conserving North America's Predators*. Washington, DC: Island Press, 2014.

Fascione, Nina, Aimee Delach, and Martin E. Smith, eds. *People and Predators: From Conflict to Coexistence*. Washington, DC: Island Press, 2004.

Feldhamer, George A., Bruce C. Thompson, and Joseph A. Chapman. *Wild Mammals of North America: Biology, Management, and Conservation*. 2nd ed. Baltimore: Johns Hopkins University Press, 2003.

Gehrt, Stanley D. *Urban Coyote Ecology and Management*. Cook Country, Illinois, Coyote Project. 2006.

Gese, Eric M. *Lines of Defense: Coping with Predators in the Rocky Mountain Region*. Utah State University Extension.

Glasgow, Vaughn L. *A Social History of the American Alligator: The Earth Trembles with His Thunder*. New York: St. Martin's Press, 1991.

Hansen, Kevin. *Bobcat: Master of Survival*. New York: Oxford University Press, 2006.

———. *Cougar: The American Lion*. Flagstaff, AZ: Northland Publishing, 1992.

Halfpenny, James C. *Scats and Tracks of North America*. Guilford, CT: Falcon Guides, 2008.

Hope, J. "Wolves and Wolf Hybrids as Pets Are Big Business but a Bad Idea." *Smithsonian*, vol. 25(3), June 1994. page 34.

Hornocker, Maurice, and Sharon Negri, eds. *Cougar: Ecology and Conservation*. Chicago: University of Chicago Press, 2010.

Hygnstrom, Scott E., Robert M. Timm, Gary E. Larson, eds. *Prevention and Control of Wildlife Damage*, 2 vols. Lincoln: University of Nebraska–Lincoln, 1994.

Imhoff, Daniel. *Farming with the Wild: Enhancing Biodiversity on Farms and Ranches*. San Francisco: Sierra Club Books, 2003.

Kays, Roland. "Yes, Eastern Coyotes Are Hybrids, but the 'Coywolf' Is Not a Thing." *The Conversation*. https://theconversation.com/yes-eastern-coyotes-are-hybrids-but-the-coywolf-is-not-a-thing-50368.

Kobalenko, Jerry. *Forest Cats of North America: Cougars, Bobcats, Lynx*. Altona, MB: Firefly Books, 1997.

Levy, Sharon. "Rise of the Coyote: The New Top Dog." *Nature*, vol. 485(7398), May 16, 2012. Pages 296–297.

Masterson, Linda. *Living with Bears Handbook*. 2nd ed. Masonville, CO: PixyJack Press, 2016.

Mastro, Lauren L, Eric M. Gese, Julie K. Young, and John A. Shivak. *Coyote (Canis latrans), 100+ Years in the East: A Literature Review*. Addendum to the Wildlife Damage and Management Conference (2012).

Mech, L. David. "Is Science in Danger of Sanctifying the Wolf?" *Biological Conservation*, vol. 150(1): 143–149. June 2012.

Mech, L. David, and Luigi Boitani, eds. *Wolves: Behavior, Ecology, and Conservation*. Chicago: University of Chicago Press, 2006.

Reid, Fiona A. *A Field Guide to Mammals of North America*. Boston: Houghton Mifflin, 2006.

Reidinger, Russell F. *Wildlife Damage Management: Prevention, Problem Solving and Conflict Resolution*. Baltimore: Johns Hopkins University Press, 2013.

Salmon, Terrell P. *Wildlife Pest Control around Gardens and Homes*. Oakland: University of California, 2006.

Shivik, John A. *The Predator Paradox: Ending the War with Wolves, Bears, Cougars, and Coyotes*. Boston: Beacon Press, 2014.

Sibley, David Allen. *The National Audubon Society The Sibley Guide to Birds*. New York: Alfred A. Knopf, 2000.

Sowka, Patricia A. *Living with Predators Resource Guide: Recreating in Bear, Wolf, and Mountain Lion Country*. Living with Wildlife Federation, 2013.

Stolzenburg, William. *Where the Wild Things Were: Life, Death, and Ecological Wreckage in a Land of Vanishing Predators*. New York: Bloomsbury, 2008.

Stone, Suzanne Asha. *Livestock and Wolves: A Guide to Nonlethal Tools and Methods to Reduce Conflicts*. Defenders of Wildlife, 2008.

Stratham, Mark J., and others. "The Origin of Recently Established Red Fox Populations in the United States: Translocations or Natural Range Expansions?" *Journal of Mammalogy*, 93(1): 52–65, 2012.

Urbigit, Cat. *Shepherds of Coyote Rocks; Public Lands, Private Herds and the Natural World*. Woodstock, VT: Countryman Press, 2012.

——. *When Man Becomes Prey: Fatal Encounters with North America's Most Feared Predators*. Guilford, CT: Lyons Press, 2014.

——. *Yellowstone Wolves: A Chronicle of the Animals, the People, and the Politics*. Blacksburg, VA: McDonald & Woodward Publishing Company, 2008.

United States Department of Agriculture. "Cattle and Calves Predator Death Loss in the United States, 2010." Fort Collins, CO: USDA–APHIS–VS–CEAH, National Animal Health Monitoring System, #643.0312, 2012.

——. "Sheep and Goats Death Loss." Report released May 27, 2010. USDA–NASS.

——. "Sheep and Lamb Predator and Nonpredator Death Loss in the United States." Fort Collins, CO: USDA–APHIS–VS–CEAH, National Animal Health Monitoring System, #721.091, 2015.

Wild, Paula. *The Cougar: Beautiful, Wild and Dangerous*. Madeira Park, BC: Douglas and McIntyre, 2013.

Wilson, Don E., and DeeAnn Reeder, eds. *Mammal Species of the World: A Taxonomic Reference., 3rd ed*. Johns Hopkins University Press, 2005.

Online Sources

Amazonarium
www.amazonarium.com.br

Animal Diversity Web
animaldiversity.org

Beartracker's Animal Tracks Den
www.bear-tracker.com

Carnivore Ecology and Conservation
www.carnivoreconservation.org

Cat Specialist Group
www.catsg.org

Centers for Disease Control and Prevention
www.cdc.gov/healthypets/pets/wildlife.html

Cornell Bird Lab
www.allaboutbirds.org

The Cougar Fund
www.cougarfund.org

Cougar Rewilding Foundation
www.easterncougar.org

Eastern Coyote/Coydog Research
www.easterncoyoteresearch.com

Eastern Puma Research Network
eprn.homestead.com

Felidae Conservation Fund
www.felidaefund.org

How to Differentiate Between Coyote and Dog Predation on Sheep
www.omafra.gov.on.ca/english/livestock/sheep/facts/coydog2.htm

Idaho Council on Industry and the Environment
www.icie.org

Interagency Grizzly Bear Committee
www.igbconline.org

International Union for Conservation of Nature
www.iucn.org

International Wolf Center
www.wolf.org

Internet Center for Wildlife Damage Management
www.icwdm.org

Journals of the Lewis and Clark Expedition
lewisandclarkjournals.unl.edu

Mexican Wolf Fund
www.mexicanwolffund.org

Mexican Wolf Recovery Program
www.fws.gov/southwest/es/mexicanwolf

Montana Fish, Wildlife & Parks
www.fwp.mt.gov

National Park Service
www.nps.org

National Trappers Association
www.nationaltrappers.com

Ontario Ministry of Agriculture, Food, and Rural Affairs
www.omafra.gov.on.ca

Raising Rabbits in Colonies
www.ecolinst.hu/upload/1/Raising–Rabbits–in–Colonies–fljaoy.pdf

Red Wolf Recovery Program
www.fws.gov/redwolf

Sheep and Goat Research Journal, vol. 19, 2004.
www.sheepusa.org/ResearchEducation_ResearchJournal_Volume19

Urban Coyote Research
www.urbancoyoteresearch.com

US Fish and Wildlife Service
www.fws.gov

Wildlife Friendly Enterprise Network
www.wildlifefriendly.org

Wildlife Hotline
www.wildlifehotline.com

Wolf Watch
www.pinedaleonline.com/wolf

The Wolverine Foundation
www.wolverinefoundation.org

Index

Page numbers in *italic* indicate photographs or illustrations.

Interior photography credits